Communication Skills
for Biosciences

Communication Skills for Biosciences

Maureen M. Dawson
Centre for Learning and Teaching
Manchester Metropolitan University
Manchester
UK

Brian A. Dawson
Former Secretary of the Faculty of Science and Engineering
The University of Manchester
Manchester
UK

Joyce A. Overfield
Division of Health Science
Manchester Metropolitan University
Manchester
UK

WILEY-BLACKWELL

A John Wiley & Sons, Ltd., Publication

This edition first published 2010, © 2010 by John Wiley & Sons Ltd.

Wiley-Blackwell is an imprint of John Wiley & Sons, formed by the merger of Wiley's global Scientific, Technical and Medical business with Blackwell Publishing.

Registered office:
John Wiley & Sons Ltd, The Atrium, Southern Gate, Chichester, West Sussex, PO19 8SQ, UK

Other Editorial Offices:
9600 Garsington Road, Oxford, OX4 2DQ, UK
111 River Street, Hoboken, NJ 07030-5774, USA

For details of our global editorial offices, for customer services and for information about how to apply for permission to reuse the copyright material in this book please see our website at www.wiley.com/wiley-blackwell

Library of Congress Cataloging-in-Publication Data

Dawson, Maureen M.
    Communication skills for biosciences / Maureen M. Dawson, Brian A. Dawson, and Joyce A. Overfield.
        p. cm.
    Includes index.
    ISBN 978-0-470-86393-0
    1. Communication in science.   2. Technical writing.   3. Life scientists–Vocational guidance.
I. Dawson, Brian A.   II. Overfield, J.   III. Title.
    Q223.D39   2010
    570.1'4–dc22

                                                                                    2009047225

ISBN: 9780470863930

A catalogue record for this book is available from the British Library.

Set in 11/13 Times by Toppan Best-set Premedia Limited.
Printed in Great Britain by CPI Antony Rowe, Chippenham, Wiltshire.

First Impression   2010

This book is dedicated to our families

Contents

Preface

It is well established that good communication skills are highly desirable in graduates in the biosciences no matter what careers they enter. Indeed, such skills are essential if undergraduates are to make the most of their degree course, since the majority of bioscience courses place emphasis on their acquisition. During their time at university, students may need to acquire a variety of presentational skills. These will typically include individual and group oral and poster presentations, and the writing of essays, practical reports, dissertations, projects, bibliographies and reference lists. In all their written, oral and visual communication, students will need to be able to summarize material while also avoiding plagiarism.

There is a perception that many students enter university without fully understanding the rules of English grammar, yet they are expected to communicate effectively in written form. Those who are skilled in written English are often unaware of what is acceptable in written scientific English. Their progress may therefore be hampered and they may become disheartened. If the basic skills are acquired at the outset, then students will gain much from their course and will emerge as more confident individuals.

This book will provide the student with practical advice on how best to communicate scientific material. It is aimed primarily at undergraduates in the biosciences, though postgraduate students may also find it useful. It will also be a useful text for students taking foundation years in sciences before tackling a degree course, as well as undergraduates on science courses generally.

Acknowledgements

We would like to thank: Lisa Coulthwaite and Sadia Chowdhury, Manchester Metropolitan University for providing posters; Len Seal for the assessment marksheets; Alan Fielding for the planning for a dissertation-based project; Sid Richards for providing the cartoons and Julia Dawson, for her hard work in literature searching while on her school placement. Many thanks also to Nicky McGirr for her patience!

1

Communication Skills in Science

About this chapter

In this chapter we will discuss the importance of communication in science and the types of communication skills you will need both during and after your university course. We will look at some basic rules to follow when writing scientific English, and at general issues, such as paragraphing, common spelling mistakes, use of the apostrophe and problems that may arise with the use of spell-checkers. Although this book is aimed principally at bioscience students, much of the advice will be useful for science students generally.

Why are communication skills important for scientists?

When successful graduates move into scientific careers, they will be called upon to practise the communication skills they have learned during their training. In addition, they may be expected to talk about their work with scientists and with non-scientists. For example, scientists specializing in the molecular biology of cancer and who work in a research laboratory may be expected to communicate their work to:

- fellow scientists working in their laboratory;
- scientists who work in different laboratories, but who may wish to collaborate;

- scientists at national and international conferences;
- eminent scientists who sit on grant-awarding authorities;
- students undertaking a placement in their laboratory;
- research students whom they may be supervising;
- journalists who want to find out about (and possibly publicize) their work;
- ethical committees, which consist of scientists and non-scientists, if their work has ethical implications;
- community groups and representatives from business who may wish to donate funds to their research;
- senior managers who may influence the future course of their work.

The types of communication skills required by today's scientists include being able to communicate in writing, and to make presentations which involve both oral and visual communication. Examples of written communications include:

- laboratory reports;
- research papers, articles and reviews for scientific journals;
- grant applications;
- briefings for management;
- progress reports;
- product descriptions.

Examples of oral communications include:

- talks to a variety of audiences;
- team or management briefings;
- research papers delivered at conferences.

Examples of visual presentations include:

- scientific posters;
- information leaflets for target audiences.

Presentations using computer software such as PowerPoint require oral and visual communication skills. Above all, it is essential that scientists communicate the results of their work in a way which takes account of the audience, but which is always truthful and unambiguous.

New students studying science at university will quickly find that they are expected to acquire and demonstrate a wide range of communication skills throughout their course. It is no longer possible for students to obtain a university degree based almost entirely on the ability to pass examinations at the end of each year, as was the case in many degree programmes fifty years ago. However, even if successful science graduates choose a career other than one in science, they will find that they require good communication skills in any 'graduate' career they enter. For this reason, communication skills are regarded as 'transferable skills' which can enhance the employability of a student in many careers.

Scientific writing: a little bit of history

Table 1.1 shows some history of scientific writing which goes back to around 1400 BC. You can see that much of the reason behind recording natural phenomena (eclipses, floods etc.) had a very practical purpose, such as being able to predict when these phenomena would occur.

Table 1.1   A brief history of early science

Date	Who and where?	What and why?
1400 BC	China Egypt	Recorded information about, for example, solar and lunar eclipses and floods in order to predict when they would occur
800 BC	Homer, Hesiod	Indicated knowledge and study of stars and constellations, probably to indicate seasons for planting crops or to provide sailors with aids to navigation
500 BC onward	Greeks	Used mathematics to lay down definitions and first principles of geometry Study of anatomy and physiology (dissection being practised) Technical terminology and taxonomy being developed
372–287 BC	Theophrastus	Produced treatises on botany; distinguished between mono- and dicotyledonous plants
ca.100 AD	Roman Empire	'Sophisticated' clinical techniques being practised

After the fall of the Roman Empire science and medicine declined in much of Western Europe but continued to flourish in the Arabic and Chinese worlds. In fact, had Arabic scholars of the period not translated much of the scientific literature of the later Roman times and of the great period of Greek science and philosophy, it is probable that such literature would not have survived. It was only when the Arabic translations of lost texts became available in Western Europe that science and scientific understanding began to be revived, and by the fifteenth and sixteenth centuries the development of science began to take off again in the Western world.

Today, we are very familiar with the idea that science is an experimental subject in which findings from experiments allow us to build on the work of previous scientists. Early on, though, science was mostly about observing rather than experimenting. The beginning of the seventeenth century, however, saw the development and widespread acceptance of what is known as the 'scientific method'. Scientific method (see Figure 1.1) involves:

An example of this might be:

1. You have observed that young pea seedlings grow towards the light but you want to test this scientifically. So, you plant several trays of seeds. One tray you grow in darkness, another in full light, another with a sole source of light which comes from one side only.

2. You provide the conditions required for growth (water, warmth, air) and, after a certain time, you measure the length of the seedling above

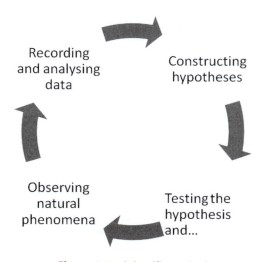

Figure 1.1   Scientific method

the soil, and note any curvature. Your analysis shows that the peas grown in the dark have grown longer than those in full light, while those illuminated on one side only have curved towards the light source. You have tested the hypothesis that peas grow towards the light, but you now need to construct a hypothesis to suggest why those grown in the dark are longer, and so on.

Scientific experimentation is a continuous process, as observations from experiments can then lead either to the support of the hypothesis and/or to the construction of new ones. It is the reporting and sharing of data that allows others to reproduce, and, therefore, to revalidate the experimental studies that to a very large extent inform the modern approach to scientific writing.

Scientific language

For quite a long time Latin was the international language of science in Europe and it was not until the seventeenth century that the use of English in scientific literature began to take off. Until then, however, science was only communicated between those who were highly educated and (usually) male. Newton's great work, *Philosophiæ Naturalis Principia Mathematica (The Mathematical Principles of Natural Philosophy)* was published in Latin in 1687.

During the seventeenth, eighteenth and even the mid-nineteenth centuries, scientific literature written in English followed the prose styles of the day and was often very conversational. For example Elie Metchnikoff, who developed the theory of phagocytosis in 1882, following experiments with starfish larvae, wrote:

> I felt so excited that I began strutting up and down the room and even went to the seashore to collect my thoughts. I said to myself that, if my supposition was true, a splinter introduced into the body of a starfish larva … should soon be surrounded by mobile cells as is to be observed in a man who runs a splinter into his finger. This was no sooner said than done.

Through the influence of the Royal Society, it became increasingly the practice to use a much simpler and more straightforward style of writing, which remains to this day, and this is what you should be aiming for. The development of scientific journals and periodicals has undoubtedly led to a much more formalized and structured approach, both in the manner that articles are set out and in the language used. This helps ensure that ideas are conveyed effectively in a way that can be easily understood by the educated reader.

Peer review

Henry Oldenburg (ca. 1619–1677) was the secretary of the Royal Society of London. He was the first editor of the Royal Society's *Philosophical Transactions,* first produced in 1665, which published the work of eminent scientists. In order to persuade scientists to publish their work, Oldenburg guaranteed that scientific 'papers' would be published rapidly, ensuring that scientists received proper attribution for their original works. He also ensured the quality and standards of the publications by sending them to experts in the field, to comment on them and to recommend publication. This process continues throughout the world in those journals which publish 'peer-reviewed' articles. A paper which has been published in a peer-reviewed journal is more highly regarded by the scientific community than one which has not, since it has been recommended by fellow scientists. However, unfortunately, on occasion, the system of peer review has sometimes delayed the publication of important works which have challenged the current view.

Basic rules for writing (scientific) English

Whether you are writing for publication in a peer-reviewed journal, for a research thesis, or for a simple laboratory report, there are certain rules you need to follow. Other chapters in this book will look at specific examples of scientific writing, such as laboratory reports, essays and so on. Here are a few rules to absorb before you get there:

1. You are not trying to write a piece of light reading or a work of fiction. You have to aim for a straightforward personal style that is understandable and readable. The material contained in your essay, report or paper must be rigorous and comprehensive. Scientific rigour may be a cliché, but it does sum up the basic overall approach. The last things you want in scientific writing are woolliness in your approach to your data and findings, and sloppiness in your use of words. Try to be precise, but do not confuse precise with brief. Use clear and unambiguous language. It is always worth remembering that if you have difficulty in understanding a passage in a textbook, then it could be that the passage is badly written.

2. Use short sentences wherever possible.
   You may have heard of the 'Plain English Campaign'. It did much to get rid of badly written English from official documents. It recommends an average sentence length of fifteen to twenty words. Wherever

possible, try to make sure that any sentence can be understood in a single read through. Overcomplicated sentence structures are totally off-putting. On the other hand, do not be afraid to vary the syntax of your sentences. A string of sentences like 'the cat sat on the mat and the mouse played in the yard' can be boring in the extreme. So, a sentence such as 'while the cat was sitting on the mat, the mouse played in the yard' is both simple and more interesting. The phrase 'while the cat was sitting on the mat' is an example of a subordinate clause, whereas the phrase 'the mouse played in the yard' is the main clause. Use subordinate clauses whenever you can, as this will allow you to bring some flowing movement into your style. Think about what you want to emphasize in your sentence. Try to ensure that the key phrase or word is not lost or split up in a mass of other words. An example of a sentence which loses its way is

> While the cat was sitting on the mat, which had recently been swept and dusted in order to keep out the dust mites, whose droppings triggered asthma in the butler, a tall and handsome man of many years, the mouse played in the yard.

The main point of the sentence (the mouse played in the yard) has been lost in all the extraneous detail!

3. Use simple words that your readers/audience will understand. Nonetheless, the use of simple words should not be at the expense of accuracy. For example, use 'rain' rather than 'precipitation', unless the context demands precipitation. Do not try to impress by using big or uncommon words where shorter words will do.

   Technical words or phrases are generally a way of conveying a complex idea in as few words as possible. The same is true of acronyms and abbreviations.[1] If you do use technical words or acronyms, explain them when they first appear. For example 'The structure of deoxyribonucleic acid (DNA) differs from that of ribonucleic acid (RNA) in …'.

   This can be particularly important in undergraduate first-year essays, as it will show your tutor that you understand the concept. For the most part, the extent to which you use technical language will depend on the readers/audience at whom your paper is aimed. Always be consistent in the technical words, acronyms and symbols you use. As far as abbreviations, as distinct from acronyms, are concerned, only use them for units of measurement.

4. Avoid colloquialisms: for example, you do not store reagents in the 'fridge' or even the 'refrigerator'; you store them at 4°C. This is because it is the temperature that is important, not the location. Similarly, where necessary, you store reagents at −20°C or −70°C, rather than in the freezer. Some chemical reactions need to take place in the absence of light. The commonly used phrase is 'stored in the dark' rather than 'in a laboratory cupboard'.

5. Avoid vague adjectives which give a poor indication of what you mean, or which could be misinterpreted. For example, 'there was a large increase' depends on what you mean by 'large'. Also, you should avoid metaphors, similes and clichés as far as possible. By the way, if you are unfamiliar with any of these terms, like 'simile', we would recommend that you use a good, concise dictionary, such as the *Compact Oxford English Dictionary*. This will also be an invaluable tool when you come to write essays. Incidentally, this particular dictionary has a very useful appendix entitled 'Effective English'.

---

[1]An acronym is a word made from the initial letters of other words, for example *UNESCO* for the *United Nations Educational, Scientific, and Cultural Organization.*

6. You should, at all costs, avoid teleology. Teleology occurs when you ascribe thoughts and purpose to objects or organisms. So, for example: 'woodlice like the dark and when faced with a choice will always prefer the dark'.

7. Wherever possible use the active form of a verb rather than the passive form. For example, 'the experiments show/showed' rather than 'it was shown by the experiments'.

8. Use straightforward verbs rather than noun/verb combinations wherever possible. For example, instead of 'reached agreement', just use 'agreed'. This is particularly important when you are writing assignments with a strict word limit, where you may be penalized for excessive word length.

9. Find out what the convention in your discipline or department is about the use of the first person, as opposed to the third person, and stick to it. One disadvantage of writing in the third person is that you constantly have to resort to the passive voice (for example, you find that you have to write sentences like 'The mouse was injected with' rather than 'I injected the mouse with'). In many university departments students are actively discouraged from using the first person 'I' and 'we' in laboratory reports. This is also the convention with many scientific journals, though others do allow its use.

   A good example of where scientists have used the first person to great effect is in the groundbreaking paper of Watson and Crick in 1953 where they proposed a structure for DNA. This paper starts off:

   > We wish to suggest a structure for the salt of deoxyribose nucleic acid (DNA). This structure has novel features which are of considerable biological interest.

10. Be consistent in your use of the tenses of verbs. When you are reporting experiments, you must use the past tense; that is, you must say what was done (or what you did). On the other hand, when you are discussing and interpreting the results and data from the experiments, you will generally use the present tense. Ensure that there is agreement between the verb and subject. This is particularly important where you are using words that have a Latin or Greek plural form (see section on plural forms). Therefore while one bacterium **is**, two bacteria **are** – this is very often misused in popular journalism. Similarly, you need to be careful with the singular and plural of mitochondrion (mitochondria) and phenomenon (phenomena). One word which appears to be changing through common use is the term 'data',

which is actually a plural word (singular 'datum'). Strictly speaking, you should say 'the data show' rather than 'the data shows'.

Wherever possible, try to keep the subject of the sentence and the verb together. To do so normally helps the logical flow of the sentence. In addition, it will help you to avoid some of the pitfalls of subject/verb agreement. If the subject of your sentence is separated from its verb by ten or twenty words, you may inadvertently use the singular form rather than the plural.

You should also avoid other mix-ups between singular and plural in your sentences, such as using a singular noun in one part of your sentence, then using a plural possessive adjective or verb when you refer to it in the second part of the sentence. For example, do not write: 'The researcher will show ... and their results indicate'; here you would use 'and the results indicate ...' or 'and his results ...' or 'and her results. ...'.

11. Use accurate punctuation, as inaccuracies can confuse the reader, and make the meaning ambiguous. One very famous example of this is the title of the book by Lynn Truss: *Eats, Shoots and Leaves*. The title arises from a joke about a panda in a bar. As we know, pandas eat shoots and leaves. This panda, according to the punctuation, first eats, then shoots a gun, and then leaves the bar. A whole section in this chapter is devoted to punctuation.

12. Plan the structure of what you are going to write. Even under examination conditions this will help you get down what you want to say. Once you have a structure, particularly for an essay or report, work on a first draft, which you can revisit as often as time allows. With word-processing this is much easier than it is with handwritten text.

13. Find out how long the work is expected to be and stick to it. If you are expected to present a 3000 word essay, do not produce something that is 2000 words long, or for that matter an epic of 10,000 words. There is much value in being concise and keeping within the guidelines given – being able to read and act on instructions is a skill in itself. Find out whether there are any prescribed guidelines on layout (for example, double-spaced text, paper size, font size and so on) and stick to them. Try to avoid giving your tutors an excuse to lower your marks.

14. If you are writing a report or an essay, it is always worth looking at the styles used in the articles in the major journals of your discipline. It is particularly useful to look at the style of the summary or

the way in which data are presented. Very often there are also standard conventions for the labelling of tables and figures, and for the layout of mathematical formulae and calculations. There is more information about the use of tables and figures in Chapters 4 and 5.

15. Adopt a logical approach to your structure. Take your arguments step by step, ensuring that the second step follows logically from the first. Adopt a clear layout, particularly, for example, if you are producing a formal report. There will, of course, be differences in the layout required for a poster presentation from those for a laboratory report or essay. Long passages of text with few paragraphs can be daunting to the reader. Think hard about the use of bullet points. They allow us to get across in summary form a number of linked ideas and are very useful in poster or Powerpoint presentations, but you need to ask yourself if it is appropriate to use them in an essay.

16. Remember that, when you first produce a piece of work such as an essay, this is only a draft – the first production is not the end product. Drafts of work can be much improved by careful reading and restructuring if necessary. Depending on your assignment, your tutor may be willing to have a look at the draft before you complete the assignment. This is particularly true of drafts of project reports or dissertations. Remember also to give yourself time to work on drafts; you should not be starting your assignment at the last minute. Working on drafts may take the mark for your work up a grade.

17. Try to put yourself in the position of your readers/audience. For the most part your writing should be pitched at the educated general reader. If you are writing a presentation at the end of your first year, ask yourself whether your fellow students will be able to understand and learn from it.

18. Always check the accuracy of any mathematical calculations or data presented in tables before you begin to revise your first draft.

19. Try reading your draft out loud, and listen to the rhythm of the sentences. If you find one that seems awkward, consider restructuring or revising it. When you think you have something close to a final draft, let someone else see it. They will not be too close to the work, and are likely to be able to spot important omissions or even grammatical or punctuation mistakes. Always run a spell-check at every stage of revision and make sure that the spell-checker is set to UK English (see below) if you are in a university in the UK or another country

requiring the use of UK English rather than USA English. Remember that a spell-checker will ignore words which, though spelt correctly, are being used incorrectly, for example, using the word 'there' instead of 'their', or 'flour' instead of 'flower'.

20. If you are using Latin words, and in some areas of biology and medicine this is unavoidable, follow the established conventions. Latin names and phrases are always written in italics or are underlined. When writing a species name in Latin the genus has a capital letter at the beginning, and the species is written in lower case. The genus may be abbreviated if it has occurred once. For example, 'the bacterium *Neisseria meningitidis* occurs in three serotypes: *N. meningitidis* Serotypes A, B and C'.

21. If you are referring to human genes it is customary to use italics for the gene, for example, the *RHD* gene encodes the Rh D antigen in red blood cells.

Punctuation

Punctuation helps the reader to understand written language and can help to avoid confusion or misunderstandings. In a sense, it is a way to help you read a sentence, or paragraph, out loud, with marks like commas indicating where you need to make a break. The following are some brief notes about the use (and common misuses) of punctuation marks.

Apostrophe

The misuse of the apostrophe is often regarded by some as a cardinal sin, and many tutors may find it extremely irritating. However, it is probably fair to say that the majority of people have misused an apostrophe at some stage in their lives. The apostrophe has two primary purposes: to denote possession and to replace missing letters in some 'shorthand' or 'contracted' terms.

Contractions

Although contracted forms are popular in speech, especially informal speech, there are relatively few contractions used in formal, written English and very few are used in scientific writing. Thus, though contractions of verbs, such as I'd for 'I would' or 'I had' or I'll for 'I will', 'I'm' for 'I am', 'isn't' for 'is not' 'we're' for 'we are' are spoken frequently, they should not be used in formal scientific writing. Conventions do change and certain words which were originally contractions have become so ingrained in our culture that the apostrophe is seldom used. For example, phone used to be written as 'phone, as a contraction of telephone. Equally, plane used to be written as 'plane, as a contraction of aeroplane.

Possession

Where the apostrophe is used to denote possession, it takes two forms: for singular nouns, it will appear as 's, for example, the cat's paw or the dog's dinner: for plural nouns it will generally appear as s': for example, the asses' ears, and the dogs' dinner (i.e. the dinner of several dogs). The exception is where there are unusual forms of plural: for example, you would use children's rather than childrens', or women's rather than womens'.

An apostrophe is never used simply to indicate a plural term and 'the dogs went to dinner' does not require an apostrophe anywhere. Thus, a greengrocer who lists his vegetables as *tomato's* or *carrot's* is incorrect. Tomato's means belonging to the tomato.

The contraction 'it's' for 'it is' gives rise to one of the commonest confusions and misuses of the apostrophe. 'It's' is **never** used to denote something belonging to it.

A word of caution: American spell-checkers may sometimes mislead on the correctness or otherwise of the apostrophe as used in the UK, so make sure your spell-checker is set to UK English.

Inevitably, language does change and develop. By long-standing convention the plural form of abbreviations does not include an apostrophe. For example the accepted plural form of DVD is DVDs and not DVD's. Equally where numbers are used in place of words, for example, the 80s, as the term for the period 1980 to 1989, no apostrophe is needed. Again if you have hundreds of books, you could use the form 100s.

Comma

The comma is used to separate parts of sentences into manageable and logical bits. It is generally used to separate subordinate clauses from the main clause, or other subordinate clauses. For example: 'The dog, which had walked all the way from the station, went to get its dinner'. However, you do not generally use a comma before 'that', as, for example, in: 'the dog that had walked all the way from the station went to get its dinner'.

In addition, with some subordinate conjunctions, like 'who', the comma can be used to distinguish between what is a necessary part of the sentence and what is there for elaboration. For example, in the sentence 'Customers who steal will be prosecuted' the subordinate clause 'who steal' is a necessary and conditional part of the sentence. In the sentence 'John, who came from London, was found wandering the streets', the subordinate clause 'who came from London' is merely an elaborative detail. It is generally conventional to use a comma before clauses beginning with 'which'. Where a subordinate clause is followed by another part of the same sentence, it is generally closed with a comma.

The comma is also used before other conjunctions like 'but', which join two main clauses. It is used after introductory words/participles: for example, 'nevertheless', 'for example', 'in general', and so on.

Other uses of the comma are: to separate a passage of direct speech from the main sentence and to separate adjectives in a list. It is used as a separator in constructions like dates, years (for example, September, 2007) or town, county (for example, Wilmslow, Cheshire). Another use is the so-called bracket comma, generally where you are linking two nouns in apposition; for example, 'His brother, William, was to become the leader of the group'.

You will normally use a comma to separate items in a list. However, if you have a long list introduced by a colon, you should use a semicolon as the separator. An example is given below:

> The aims of this project are as follows:
>
> 1. to investigate the effect of temperature on enzyme activity;
> 2. to investigate the effect of pH on enzyme activity;
> 3. to determine the K_m of an enzyme.

The commonest misuse of the comma is in the so-called comma splice. This is where a writer uses a comma instead of a full stop or semicolon where there are two separate main clauses without a conjunction. For example: 'he went to the shops, the streets were bare'. In these circumstances, you should always use a full stop or semicolon.

Colon

The colon is used as a mark to introduce a list, and each item in the list is separated by a semicolon. By convention it is also used with bullet points, and, again, each bullet point is separated from the next by a semicolon. The colon is also commonly used to introduce a linked explanation or expansion of the clause before it; for example, 'the one fact you should remember about chilli powder: it is hot'. A colon here represents a slightly more emphatic break than a comma. For this reason it is sometimes used to emphasize the following word or phrase; for example: 'she had one great discovery: radium'.

Dash

Do not use a dash to replace a comma, unless the use of a comma would be ambiguous, as in a list where you need to put something in parenthesis; for example, 'his two brothers – Jack and James – his father and his uncles...'. It may be more appropriate here to use parentheses (brackets).

Exclamation mark

As may well be expected, the exclamation mark is used to indicate surprise, astonishment: 'Surprise! Surprise!'. Try to avoid using exclamation marks just to add emphasis. They should rarely be used in scientific writing, but can add interest if used appropriately when writing for the general public.

Full stop

A full stop is generally used at the end of a sentence. An exception to this rule is where you have complete sentences in bullet points. At the end of each bullet point you should use a semicolon, with the exception of the final sentence of the final bullet point, where you will use a full stop.

Full stops are traditionally used with abbreviations, as with shortened species names such as *E. coli*. Full stops are not used with acronyms. Thus we write 'AIDS' not 'A.I.D.S' for acquired immunodeficiency disease.

Hyphen

The hyphen is most commonly used in compound words such as 'long-standing', and always with compounds that begin with: self, mid, all, ex, over and so on. You will also use it with part words; for example, 'you might want to check the full list of compounds beginning with fluoro-'. The hyphen is useful because it helps to avoid ambiguities. So, for example, the sentence 'the T lymphocyte kills the virus infected cells' could be confusing until the hyphen is added to make it clear that the lymphocyte is killing the virus-infected cells. Similarly the phrases 'drink induced disorder' and 'drink-induced disorder' have subtly different meanings. Unfortunately, the trend in publishing increasingly seems to be to leave hyphens out and this can lead to ambiguities and compound words strung together.

Parentheses (brackets)

Brackets can be useful in flagging up an alternative word or phrase, such as 'parentheses (brackets)' or to indicate an abbreviation such as deoxyribonucleic acid (DNA). It is possible to overuse brackets and it is always worth considering whether commas could do the same job.

If you include a full sentence in parentheses, you will normally include the full stop before the final parenthesis.

Question mark

The question mark is self-explanatory. You can use it when asking real or rhetorical questions even though, with the latter, you are not expecting an answer.

Quotation marks (inverted commas)

Quotation marks are traditionally used to mark the beginning or end of a quoted passage. However, there is a growing convention that, for other than short quotations, italicized text indented from the main text is used. It is also commonly the case that italicized text, without indentation, is used for shorter passages. In the end it is a matter of style. Whichever approach you use, be consistent throughout your text, but also bear in mind that people who suffer from dyslexia may have difficulty reading italicized text. In this book, we have chosen to use quotation marks and normal text.

If you look at a computer keyboard you will see that there are two options for quotation marks: the double ("…") or the single ('…'). In general, they are used interchangeably. However, where you do get a quotation within a quotation, the convention is to use the single form to mark the beginning and the end of the main quotation, and the double form to mark the end of the quoted quotation. Again, be consistent in your use throughout your text.

Quotation marks are sometimes used to mark titles (for example, of books or articles). In general, italics are used in scientific literature.

Semicolon

In many ways the use of the semicolon overlaps that of the comma, but generally it is seen as indicating a more significant gap. The most common use of the semicolon is to separate two closely related statements that could each form a separate sentence. In these circumstances you cannot replace the semicolon with a comma. It is also used to separate items in a long list after a colon and in bullet points. As we have already seen, at the end of the final sentence of each bullet point you should use a semicolon, apart from the final sentence of the final bullet point, where you will use a full stop.

More on plural words

Many of the words used in science in general and the biosciences in particular are either derived from or are actually Greek or Latin words. This is not really surprising, as many of the major figures in science or the biosciences would have had Latin and Greek as part of their education, particularly those who lived in the nineteenth century or earlier. Very often writers may not always be aware that some words they are using are plurals. Even those guardians of English, the British Broadcasting Corporation (BBC), get it wrong on occasions. The commonest errors relate to the plurals of neuter words in Greek or Latin, which have an ending in -a. The four words that confuse people the most are: bacteria (singular bacterium); data (singular datum); media (singular medium) and phenomena (singular phenomenon). So, you should never say for example, 'this bacteria is', or 'this data is': it should always be 'these bacteria are' or 'these data are'. Other plural words include mitochondria (singular mitochondrion) and spermatozoa (singular spermatozoon).

In many cases the Greek or Latin plural form is retained. For example the plural of stoma (the epidermal pore in plant leaves) is stomata. In other cases, either the Greek or Latin plural form or an English plural form maybe used; for example, the plural of octahedron is octahedra but octahedrons is also acceptable. In other cases, only a normal English form is used: for example, the plural of virus is viruses. There are also several words that appear to have a Latin or Greek singular form. For example the plural of octopus is octopuses, similarly the plural of platypus is platypuses, and the plural of polygon is polygons. The reason for this is that the -pus element of octopus is derived from the Greek word for foot, and the -gon element of polygon is derived from the Greek word for corner.

The best approach is to try to remember those words with a confusing singular and plural.

Commonly confused words

There are many words that are commonly confused with others. In many cases they are homophones (words pronounced in the same way but having a different meaning) or homonyms (words spelt in the same way, but having a different meaning). Here are some of the pairs of words that you are likely to come across.

Accept and **except**: 'accept' means to receive; 'except' means (as a verb) to omit, or, as a preposition, it means 'not including'. Example: 'I accept all the terms of your offer except the one requiring me to jump off the cliff'.

Affect and **effect**: 'affect' means to influence; 'effect' (as a verb) means to cause to occur.

Albumin and **albumen**: 'albumin' is a plasma protein, whilst 'albumen' refers to the white of an egg.

Appraise and **apprise**: 'appraise' means to assess the worth of; 'apprise' means to inform.

Aural and **oral:** 'aural' refers to the ears, while 'oral' refers to the mouth.

Complement and **compliment**: 'complement' means the part that completes something, or, in the biological sense, a group of proteins in serum that destroys cells when activated by antibodies; 'compliment' means a polite expression. Similarly, you should not confuse complementary with complimentary.

Continuous and **continual**: 'continuous' means without a break; 'continual' means recurring at regular intervals.

Council and **counsel**: a 'council' is an administrative body; to 'counsel' as a verb means to advise. As a noun 'counsel' is a legal adviser.

Draft and **draught**: you 'draft' a paper, but get cold when you stand in a 'draught'.

Defuse and **diffuse**: 'to defuse' means take the fuse out of (for example, a bomb) or remove the cause of tension; 'to diffuse' means to scatter over a wide area.

Discreet and **discrete**: 'discreet' means tactful or unobtrusive; 'discrete' means separate.

Formally and **formerly**: 'formally' means conventionally; 'formerly' means previously.

Imply and **infer**: 'imply' means to suggest or insinuate; 'infer' means to conclude or deduce.

It's and **its**: 'it's' is the contracted form of it is or it has; 'its' means belonging to it (see section on the (mis-) use of the apostrophe).

Lead and **led**: 'lead' pronounced with a short e sound is the metallic element; 'lead' with a long e sound is the present tense of the verb 'to lead'; its past tense form is 'led'. You should never use lead, if you want to express the past tense of the verb 'to lead'.

Lay and **lie**: 'lay' can only be used with an object; 'lie' can only be used without one. For example, he lays a cloth on the table; he lies on the bed.

Lightening and **lightning**: 'lightening' means making light or less heavy; 'lightning' is a flash of light in a thunderstorm.

Loose and **lose**: 'loose' (as a verb) means to release or set free; 'lose' means to cease to have.

Mitigate and **militate**: to 'mitigate' means to make something less severe; to 'militate' is to work against something.

Plural and **pleural**: 'plural' refers to more than one; 'pleural' refers to the lungs.

Past and **passed**: 'past' means completed or finished; 'passed' is the past participle of the verb to pass.

Practice and **practise**: 'practice' is the noun, and 'practise' is the verb. An easy way to remember the difference is to think of advice, and its corresponding verb advise.

Principal and **principle**: 'principal' (as an adjective) means first importance or main; principal as a noun refers to a head of an institution such as a college; 'principle' is a noun normally meaning a fundamental or general truth. For example, 'The principal reason I am here is to set out the principles of biology'.

Role and **roll**: 'role' is a part or character in a play; 'roll' has a great many meanings, including to move by turning, a rounded mass, a cake of bread etc.

Stationary and **stationery**: 'stationary' is an adjective meaning not moving; 'stationery' is a noun meaning paper or any writing materials.

There, their and **they're**: 'there' is an adverb meaning at that place; 'their' is an adjective meaning belonging to them: 'they're' is the contracted form of they are.

To and **too**: 'to' is a preposition generally used to indicate direction or to mark the infinitive of a verb; 'too' is an adverb meaning either as well or extremely. A nice example of the differences can be seen in the sentence: he was too ill to go to work.

Weather and **whether**: 'weather' refers to meteorological conditions; 'whether' is which of the two.

Who's and **whose**: 'who's' is the contracted form of who is or who has; 'whose' generally means belonging to who.

Your and **you're**: 'your' means belonging to you; 'you're' is a contracted form of you are. A nice example of use is: you're your own worst enemy.

If you work on the principle that you will not use abbreviations like 'they're' for 'they are' or 'it's' for 'it is', you are half way to solving some of the most common confusions in English. 'There's' will never be confused with 'theirs'.

Commonly misspelled words

Nowadays, when most assignments are required to be word-processed, it is far easier to check your spelling before your assignment is handed in. Tutors will be very sympathetic about the problems of students with dyslexia, and will make reasonable adjustments to accommodate and recognize their difficulties. However, amongst non-dyslexic students there are few things that irritate tutors, examiners and supervisors more than essays or presentations that contain misspellings and typographical errors as these usually indicate poor, or no, proof-reading. Occasional mistakes under examination conditions are understandable. However, where an essay is presented as part of the written coursework of your programme, or where materials for overhead projection or a poster have been prepared for a presentation, to neglect to use a spell-checker is to throw away marks unnecessarily. Table 1.2 lists a few commonly misspelled words. As you will see, some of the words crop up in the list of commonly confused words.

Table 1.2   Some common spelling errors

Correct	Note
Accommodate	Two 'c' and two 'm'
Address	Double 'd'
Amend	Only one 'm'
Business	An 'i' between the 's' and 'n'
Committee	Double 'm', a double 't' and a double 'e'
Controversy	An o in the middle
Definite	Spelt with -ite at the end
Fluoresce, fluorescent and fluorescence	Fluor- not flour- or flor-
Independent	end in 'ent' not 'ant'
Inoculate	One 'n' not two
Liaise	Two 'i'
Licence and license	Licence is the noun; license is the verb
Millennium	Double 'l' and a double 'n'
Miniature	Do not forget the 'a' after the mini
Necessary	One 'c' and a double 's'
Practical	Spelt with a -cal at the end not -cle
Rhythm	One of those commonly misspelled words, like phlegm, derived from Greek, that are best committed to memory
Withhold	And its related forms (like withholding) have a double 'h'

UK and American English

The spellings in this book are in UK (Oxford) English, which is the publisher's requested style. American English is often different and spellings may vary. This is worth knowing since many publishers are now multinational and may choose to produce all books in American English. However, journals based in the UK will require UK English. It is always best to find out before you begin writing.

When you are writing your assignments in the UK it is important for you to remember that UK English is required. Note also that most personal computers, particularly those which use the Windows operating system, have American English as the default. It is possible to change to UK English but if you forget, the spell-check may indicate that an English spelling is incorrect. Some common differences are shown in Table 1.3.

Note also that many English words which end in '-ise', such as 'advise' and 'surprise' end in '-ize' in American English. Also, you may find some English words ending in 'ise' or 'ize', such as 'recognise' and 'recognize'. The forms we use in this book are those given in the *Oxford English Dictionary*. If in doubt, use this dictionary as your source of reference. Also, in UK English, the

Table 1.3 Some differences between American and UK English

UK English	American English
anaemia	anemia
centre	center
chequered	checkered
colour	color
defence	defense
foetus	fetus
glycaemia	glycemia
haemoglobin	hemoglobin
humour	humor
licence (noun)	license (noun)
manoeuvre	maneuvre
metre	meter
oedema	edema
oestrus	estrus
programme (except a computer program)	program
skilful	skillful
speciality	specialty
sulphur	sulfur
traveller	traveler
tyre	tire

noun 'practice' is distinguished from the verb 'to practise' by its spelling – in American English, they are both spelt as 'practise'.

Further reading

Barass, R. (2002) *Study! A Guide to Effective Learning, Revision and Examination Technique*. London: Chapman & Hall.

Collins Essential English Dictionary, 2nd edition (2006) Glasgow: HarperCollins.

Committee on Science, Engineering and Public Policy (1995) *On Being a Scientist: Responsible Conduct in Research*, 2nd edition. London: The National Academies Press.

Compact Oxford English Dictionary, 3rd edition (2005) Oxford: Oxford University Press.

Metchnikoff, E., quoted in F. Heynick (2002) *Jews and Medicine: An Epic Saga*. KTAV Publishing House.

Northedge, R.A. (2005) *Good Study Guide*. Milton Keynes: Open University Press.

Parkinson, J. (2007) *i before e (except after c)*. London: Michael O'Mara.

Trask, R.L. (1997) *The Penguin Guide to Punctuation*. London: Penguin.

Truss, L. (2005) *Eats, Shoots and Leaves: the Zero Tolerance Approach to Punctuation*. London: Profile.

Watson, J.D. and Crick, F.H.C. (1953) Molecular structure of nucleic acids: a structure for deoxyribose nucleic acid. *Nature*, **171**, 737–38.

2

Using Scientific Literature

About this chapter

In this chapter we examine what we mean by scientific literature and the significance of peer review. We also consider why using scientific literature is essential for your essay or project work. We will examine issues concerning the sourcing of information: for example, books, academic journals, search engines and databases. We will highlight the pitfalls of using web-based information and suggest ways in which you can evaluate source material and its appropriateness to your task. We give guidance on how to read the literature critically and effectively, and how to summarize information clearly. We will take a look at plagiarism, particularly what is covered by the term, and how you can avoid it. Finally, we discuss the accepted ways in which the work of others is acknowledged, and standard styles for citing in your text and listing references or providing a bibliography.

What is scientific literature?

Scientific literature is the storehouse or archive of the history of scientific research and the techniques that researchers have developed or used. As such it is central to the development and exploitation of science itself. All scientists draw on scientific literature both to set the context of their own research and to inform fellow scientists (their peers) about the way in which their theories have developed from earlier research, and the extent to which they complement or contradict previous findings. Scientists often repeat the experiments of others, both to check that the techniques and the findings of the work reported are valid

Communication Skills for Biosciences   Maureen M. Dawson, Brian A. Dawson and Joyce A. Overfield
Copyright © 2010, John Wiley & Sons Ltd.

and to improve their own techniques. As a result the outcomes of research are constantly checked and validated and new interpretations and experimental techniques developed. It is normal for scientists to describe how a particular technique was developed and why they used it in their experiments. There is also an unwritten expectation that they will describe fully what results they obtained, how they were derived and the reasons that they have put a particular interpretation on them.

A little bit of history

Scientific literature goes back a long way. Both the Chinese and Babylonians recorded information about the eclipses of the sun and moon and other astronomical events more than two and a half thousand years ago, largely for religious reasons. It is also thought that cave paintings of animals and the like were a way of recording information for others. The Egyptians recorded the changes in the flow of the Nile for agricultural purposes. Greek and Roman authors such as Aristotle, Theophrastus and Pliny contributed significantly to the understanding and interpretation of nature and natural events. By the Middle Ages Arabic scholars were routinely undertaking experimental work and recording their results.

The development of scientific journals, as opposed to single-authored books and treatises, began in Europe in the mid-seventeenth century. An early form of peer review was introduced by the editor of the *Philosophical Transactions* of the Royal Society of London. Inevitably, as research and techniques developed, scientists found it increasingly difficult to keep up to date with the work of others, and journals began to include abstracts of current publications to inform their readers. By the beginning of the twentieth century there were formal indices or abstracts of published work and by the mid-twentieth century volumes like the *Science Citation Index* were routinely published. Scientists are constantly updating their knowledge and techniques by reading the literature in their discipline and passing on to others details of their own work. Nowadays as a result of developments in computing it is possible to bring together data from a very large range of published sources in meta-analyses. These can be important in clinical medicine, particularly when they are able to identify significant developments from the results of a large number of clinical trials that may not be entirely clear from individual case studies.

Use up-to-date sources of information

Interpretations of results do change over the years and, with hindsight, some earlier interpretations have sometimes been shown to be inaccurate or misleading. There are many such examples in the field of palaeontology, where scientists have had access to a much wider range of fossil remains, and where dating techniques have become progressively more accurate. You can think of scientific literature as a dynamic entity: there will be mistakes and misinterpretations, but these will come to light as research in the field of study develops.

Nowadays many journals are available online, and subscribers are informed in advance, by email, about the contents of the latest edition. Equally, current work is often publicized through presentations at conferences or even local research meetings where draft results are made available. Unfortunately the cost of journal subscriptions can be substantial, and sometimes even major university libraries cannot afford the subscriptions to the full range of journals available. However, many scientists get round this by collaborating closely with colleagues in other institutions or research groups that have access to the publication in question. Lack of availability of journals can represent a major difficulty for distance-learning students who do not have ready access to a university library. It is worth bearing in mind that where you do find key articles that are not available in your library, or where your nearest source is a local general library, you can get books or copies of journal articles through the inter-library loans scheme.

Today scientific literature is largely written in English, even where the first language of the authors and research groups is not English. Where you do come across an article in another language that you need to read for your assignment, you will find that many libraries offer translation facilities. In addition, you may well find translations available on the web.

Different sources of scientific literature

Scientific literature is divided broadly into two categories: primary sources and secondary sources.

Primary sources include:

- research papers in peer-reviewed journals;
- research monographs;
- conference proceedings;
- doctoral theses kept in university libraries.

Secondary sources include:

- textbooks;
- edited books;
- review articles.

Primary sources

Papers published in peer-reviewed journals are primary sources of research information, because they set out research methodologies and results. The peer-review process and the date of publication of the journal will mean that an article is published several months after the original typescript was submitted.

Research monographs, which are highly detailed pieces of scholarly work, devoted to a single topic, may also be primary sources, as they bring together authors working in the area to write papers on their research. A research monograph is usually of book length (around 50,000 words), is written for a specialized audience and generally describes leading-edge research. Research monographs in science are often multi-authored, especially where they relate to the work of a major international or collaborative group such as researchers working with the Large Hadron Collider (LHC) at the Conseil Européen pour la Recherche Nucléaire (European Council for Nuclear Research), generally referred to as CERN.

Monographs often include a survey of the latest research drawn from a range of sources, including recent conference proceedings. As such they normally go beyond the author's own research, in order to set the work in the context of recent developments. They also include some kind of historical review to show how the author has built on and extended, or even contradicted, earlier work, and how the author has contributed to the development of the subject. Monographs use technical terms as a matter of course, and assume that the reader will understand them. Very often the monograph will be part of a published series that includes what are regarded as the key books in the discipline aimed at the professional scientist. Finally, a research monograph often entails a much longer period of preparation: it takes longer to write, and involves the writer in consulting the most recent journal articles and conference proceedings. Where you are designing a research project, you would normally rely much more on research monographs and draw on a wider range of journal articles and conference papers than you would for an essay.

In addition to journal articles and research monographs, conference proceedings, which contain details of the papers and posters presented at a conference, are also primary sources as they report the original work of the authors concerned. Equally, a doctoral thesis or a bound collection of papers submitted in support of an application for a higher doctorate, such as Doctor of Science (DSc), are also regarded as primary sources. These are not generally so widely available, and copies would normally only be available in the library of the university that awarded the degree, or in the department or school of the student. You may find that your tutor or supervisor has also supervised the student concerned, and that they have a copy of the thesis. The British Library also keeps copies of the abstracts of successful PhD theses across the UK.

The key features of primary sources in the sciences are that they are the original source of a set of data, or the first report of particular observations. Sometimes a report from a government department or other organization, such as a report about the decline in the population of a particular species of birds, is seen as a primary source. It may even be the case that information available on podcasts or blogs may be regarded as a primary source, especially where the researcher wants to make his or her results available quickly to the scientific community. However you need to be cautious here, as the results and techniques described may not have been subject to peer review.

Secondary sources

Secondary sources draw on the primary sources, and include textbooks, edited books and review articles. Some, such as articles in encyclopaedias, are of limited value and may only be useful when you are looking up a term that you have not come across before. A comparison of a textbook or chapter in a textbook with a research monograph or journal article highlights the key differences between primary and secondary sources. Textbooks cover broad topics and are aimed at different levels. Some will be for first-year undergraduates or students in the later stages of secondary education. Textbooks are likely to be relatively straightforward in their language, and will avoid too many technical terms. Unlike research papers and monographs, textbooks will generally include a glossary, which will give an explanation in simple language of the technical terms commonly used in your subject or discipline. Very often textbooks aimed at first-year undergraduates will contain exercises at the end of each chapter, with model answers at the end of the book. These actually form a very useful means of checking whether you have understood what the chapter covers. For these reasons, they may also appeal to the general reader who is interested in the subject.

Other textbooks are aimed at graduate students or final-year undergraduates. These are likely to include much more technical language, and assume a higher level of background knowledge.

Some authors who are very distinguished in their research field will write books aimed at the general reader. Professor Stephen Hawking's *A Brief History of Time* is a well-known and widely read example. Another useful series is the 'Very Short Introductions' published by the Oxford University Press. Such books tend to avoid technical language or the use of formulae, particularly where they are not readily understandable. Instead, the author will set out the concepts in simple and straightforward language. As such, they often provide a good example of the sort of language and approach that you should use in writing general essays. By contrast, a textbook will normally include the essential technical terms, and will define them where they first occur, as well as in a glossary. This definition will describe the terms or concepts in simple language and will try to relate them to everyday ideas and language.

Where to start

The normal starting point for an essay will be your lecture notes and handouts and a recommended textbook. From these you should move to review articles, followed by research papers. A good textbook or review article will set the historical background, identify current ideas and set out where issues remain to be solved. Sometimes review articles will be derived from keynote presentations at major conferences, and will generally give an excellent overview of current research in a straightforward way.

Journals are often quite focused in the subject that they cover. Others, such as *Nature*, will cover a wide range of topics. Equally the standards of journals will vary, and there are published ratings of journals. It is likely that the researcher who uses novel and innovative techniques for his or her work, or who has come up with results that may change thinking in the subject, will try to have the article published in the most prestigious journal.

Publications in journal articles and research monographs are often used in recruitment and promotion procedures by universities for academic and research staff. In addition, in order to obtain research funding, academics must have a good record of publication in prestigious journals. Consequently academics will aim to publish regularly and therefore information about up-to-date research is now more readily available than in earlier times. For example, during his lifetime Mendel did not publish much of his research observations on hybridization and variation in plants, and where he did they were in fairly remote journals like the *Proceedings of the Natural History Society of Brünn* and, of course, were written in German. Consequently Darwin was not aware of Mendel's research, and it was not until the twentieth century that the real significance of this research was recognized.

For the most part it is relatively easy to follow up research in individual disciplines. Where multidisciplinary research is involved, following up information and results may be less straightforward. Publications are likely to be in a range of journals across the disciplines concerned. Researchers in the single disciplines may not pick up the work published by multidisciplinary groups, and in some instances may dismiss it as irrelevant. However, multidisciplinary groups may well bring a new and important approach that conventional single groups have not spotted. A clear example in the biosciences is where there is research that involves neuroscientists and immunologists; a comparable example in the clinical sciences is where clinical psychologists have input to the management of disease.

Peer review

Peer review is generally seen as the most effective way of maintaining the standards and quality of research, as well as ensuring that the public in general have confidence in what scientists are doing. The nature of peer review has developed as the dissemination of the literature has improved. In the seventeenth century articles published in the *Philosophical Transactions* of the Royal Society of London were reviewed by the editor, who had sole responsibility for the quality of the article published. By the middle of the eighteenth century the Royal Society itself took over responsibility for the review process, and each article was sent out for review by a limited number of experts in the field. This process was adopted increasingly in the nineteenth century, and by the twentieth century it became absolutely standard. Peer-reviewed papers, such as journal articles, are seen as the gold standard of scientific literature.

Today, an original manuscript submitted to a peer-reviewed journal is reviewed by at least two independent scientists. The reviewers must not be directly involved in the research described in the article, but will be acknowledged experts in the field with wide experience of the research methodology used and the interpretation of the results derived from the experiments described. In some cases they may be members of another team that is carrying out similar research, and the work of their team may well be in direct competition with that presented in the article. Generally the reviewers have the option of remaining anonymous. They are also under an obligation to keep information about the work confidential while they are reviewing it and until it is published. It is normal for reviewers to give feedback to the author through the editor, with the aim of improving the quality of the article. Reviewers will look at the research methodologies described, the data presented and the interpretation of the data. If they do not think that it is up to scratch, they will recommend that it should not be published in its present form, or in some cases not even published at all. They will often suggest modifications that the author should incorporate before the article is published. Anonymity can help ensure that reviews that are critical of the research are provided. Equally, of course, anonymity can be abused. The editor of the journal will look at all reviewers' reports. If there are disagreements between the reviews, they will normally seek the views of another reviewer.

Over the years there have been some concerns about the peer-review process and its effects on the publication of research. For example, peer reviewers may be reluctant to support the publication of negative results, or results that show no overall trend. One consequence of this is that the scientific literature may be skewed by the publication of largely positive results and you should bear this in mind when you are researching the literature. This can have a significant impact on the meta-analysis of the data that form the basis of the research results. There is also an additional concern that the reviewer may take a more positive view of research submitted by a prestigious university or research institute as against that of a junior academic or researcher in an institution that has a poor track record for research. However, it is worth noting that the research carried out by junior researchers, by doctoral students or by undergraduates undertaking project work can make a significant contribution to science. An undergraduate who has a peer-reviewed publication arising from project work is praiseworthy indeed, and there are even undergraduate research journals for students to publish their work. One such journal is *Bioscience Horizons*, which is a national journal for undergraduate research.

Beyond peer review

Where a new research technique has been published in a research journal, other scientists will try to reproduce the techniques and associated results, to check that both the method described is reliable and that similar results can be obtained in a different laboratory. In this way the peer-review process continues beyond the initial publication of an article in the form of post-publication replication of experiments and results by the wider research community. There have been many instances where scientists have not been able to reproduce the results. A well-known example is cold fusion, where other researchers were unable to reproduce the results described by Fleischman and Pons (1989).

Why use scientific literature for your assignments?

The development of the ability to put over ideas about your chosen area of study, either by means of oral presentation or through essays and other written work, is a vital part of your education. It is also one of those skills that can be used later in your career, whether you become a professional bioscientist or not. The writing of essays or preparation of presentations during your course is intended to help you develop a structured approach to finding and evaluating information, and setting out and communicating this information to your readers or audience. Normally the development of these skills will take time, and practice makes perfect. Moreover, the tricks and techniques acquired in the process will help you in later life to look critically at articles you may read in the press about, say, the latest 'miracle cure' for some disease or other. In addition, by successfully communicating the results of your work, you will be adding to the pool of knowledge in your area of study, and what you say may very well help others in their understanding of the field.

Your department or school will have conventions and guidelines to help you to prepare your written work or your oral presentation, and you should make sure on every occasion that you find out what these are. Many of these will be concerned with the way in which your material is presented. In your presentations, whether written or oral, you will be expected to cross-refer to the literature that you have consulted during the course of your work. In all scientific and technical writing there are established conventions that you have to follow when you are reporting or acknowledging sources of information. As much as anything, these are to ensure that it is clear what your own work is, and what is the work of others. This is not intended to prevent you from drawing on and making use of the work of other people. Indeed, as we have already said, much scientific work is the result of collaborative effort, or builds upon the work that others have undertaken in the past. Citing or cross-referring to the work of others helps strengthen the perception of the quality and breadth of your work. This is particularly so when you are drawing on the primary literature in peer-reviewed academic journals or research monographs, where the authors have reported the information for the first time. As we said earlier, textbooks are regarded as secondary literature, as they reflect their authors' views on and interpretations of the primary sources. You should remember that there may be other interpretations than those set out in the textbook you are using. However, a well-written textbook will summarize the range of interpretations and will therefore be invaluable for essays in the early years of your course. In addition, textbooks will usually provide a bibliography setting out references to the primary sources.

Newspapers and popular magazines, and to some extent encyclopaedias, are not generally useful sources. In the case of newspapers and popular magazines, the journalist responsible for the article may very well have topped and tailed a press release, which in any case has probably not been written by the original scientist.

Reading the literature

The assignments you are given during your course will inevitably require you to consult the relevant scientific literature, and without help and guidance this can seem to be quite a daunting task. As part of your A-level or comparable level work you may well have undertaken some kind of project or coursework, and will therefore have had some experience of finding out information from the Internet or from textbooks. As you will have discovered, reading scientific literature is very different from reading a novel or a magazine article for pleasure.

When you read a novel, you will quite often skim-read or skip paragraphs, and you will certainly not try to analyse it in any depth. By contrast, when you read scientific literature as part of your course, you must have a very focused approach. It is important that you read critically and think about what the author says. You should also look at the way the author presents his views and the data that support it. Equally, you should aim to understand why the book or article is seen as important. With such a focused approach you can develop your own style by learning and drawing on the stylistic techniques that an author uses. This means that reading and understanding a textbook or research article will take much longer than reading a novel or magazine article of corresponding length.

Learning to read scientific literature critically is an important skill that you must acquire and develop as your course progresses. It is a skill that you need to practise constantly. If you work on the basis that you will take two minutes to read the average page, then reading a 360-page book will take at minimum twelve hours of solid reading. If you have only, say, an average of two hours available a day for reading, it will take you the best part of a week to read the book. This, of course, does not include any time for you to take notes. You need, therefore, to focus clearly on what you have to read, and if it is obvious that an article or chapter is peripheral to your essay or project, ignore it, at least as part of your initial reading.

You should also take account of your deadlines. If you have a month to complete your essay or project, you can spend more time on reading than if you only have a week. The aim of this section is to give you some tips that will help you to develop a systematic and methodological approach to reading.

1. Check the reading list

When you start a unit or module, you will be given a reading list by your tutor. You may well be dismayed at the amount of reading suggested, especially when you realize that this is the reading list for just one small part of the course. As we said earlier, the starting point for any essay or project work is always the lecture notes and a good textbook. You should not rely just on your lecture notes and handouts. You have to read much more widely if you are going to produce a good essay, or design a good research project. Staff in your department or school will have identified the most appropriate textbook for your module or course. In some cases they may even recommend a very general textbook for your subject, for example, a standard textbook on zoology, if that is your main subject.

2. Identify your source material

Read and re-read the essay title or the subject of your project, and make sure that you understand it fully. In this way you will be able to identify the sort of source material such as notes, textbook(s), reviews or journal papers, and the information that you will need for your essay or project.

3. Think about your approach to reading

Before you start to read a book or an article, you need to think about what your approach to reading is going to be. If you are reading a textbook to get an idea of the context of your essay subject, your approach will certainly be different from that which you will use when you read a journal article as part of the background for a research project. For a research project you will need to concentrate hard on the detailed experimental techniques and the data that form the basis of the conclusion of the article.

4. Start with the textbook and use all its facilities

Textbooks are immensely helpful for identifying the key researchers in the subject, and the best primary sources of information for you to follow up when you read further into the subject. Textbooks essentially take the form of a review and summary of the subject based on the important primary sources. They will generally reflect the opinions of their authors and sometimes their prejudices. Consult the glossary for any technical terms you have not come across before. Use the bibliography to find the important primary sources: these will form the background reading that you will need to do when you are researching information. You will often find that there is much in common between the bibliography and the reading list for your module or project and that will give you a good starting point in identifying the main authors and papers.

You should also read the textbook alongside the handouts and lecture notes. You may find it helpful to highlight the key points in your lecture notes, and where possible make a note of related passages or sections in the textbook. There will be a lot more detail in the textbook, and you will find it helpful for your later revision to summarize the important parts as bullet points to supplement your lecture notes. In this way you will get into the habit of reading around to add to the information that your lecturer or tutor gives, and gain a much more informed grasp of your subject. You will also find that reading a textbook and lecture notes will be somewhat different from reading the primary literature of your subject.

5. Read the type of literature appropriate to the assignment and use an interpretive approach

When it comes to reading as part of the background research for your essay or project, you must have a very structured and disciplined approach to ensure that you are able to recall the key points at a later date. The topic of an essay may be a broad general review, or may require you to focus on a specific topic. In general, your tutor is much more likely to be impressed by a well-argued personal interpretation than by a simple narration of the facts. You will normally have to use such an interpretative approach later for, for example, a research project.

A project will almost certainly be based around a specific topic. Here you will be required to present the data that come from your research in a clear and logical way, but also interpret the implications of the data in setting out your results.

Again, the starting point will be your lecture notes and the textbook, but now you will also be expected to read much more of the primary literature, with the aim of developing your own thoughts, rather than just regurgitating the points that you have picked up from your lecture notes or the recommended textbook.

6. When reading a textbook

Inevitably, when you are writing a general essay you will have to draw initially on a lot of basic subject information from your textbook and lecture notes. Summarize the major issues as bullet points in your notes, and then follow up each bullet point by reading the main primary sources of material, usually journal papers or reviews. You should summarize the more detailed information from the papers as sub-bullets.

7. When reading a research article

If the topic of your essay is more specific, then the focus for your research may very well be a particular research monograph by a leading scientist that deals with the topic. Again, for both a general and specific essay you will need to follow it up by reading appropriate journal articles. This is also the approach that you should adopt for project work, which is much more likely to focus on a specific topic. For example, if you are writing an essay on something as broad as the causes of disease, you will need to read a general book on the topic, and then look at a limited number of specific diseases to identify common factors, and examples of marked differences, to allow you to put forward a personal view. By contrast, if you are writing an essay or doing a project on cytokines, you would need to read one or more books that are specific to the area, and follow them up by reading the key research papers from journals.

The approach to reading journal articles will have some things in common with reading a book, but will also have some differences. A book is much longer for a start (on average around 60,000 words; a journal article may be around 5000 words), but this does not mean that reading a research article is easy. A journal article will usually have an abstract, a formal introduction, followed by a description of the experimental methods and, where appropriate, the materials, that have been used, together with the results obtained. It will also include a general discussion and a conclusion, as well as a list of references or a bibliography. An article is generally more condensed and technical in its language than a book, especially a textbook, and will not normally include a glossary of terms used.

When you have found an article or paper that you think is central to the topic of your essay or project, you will need to read it more than once to ensure that you fully understand it. A good approach is to read it once, then to read a number of complementary articles or papers that have been mentioned in the text, particularly where they help in setting the context for the original paper. You will find that when you read the original article or paper for a second time, you will have a much better understanding of what its author is setting out to demonstrate, and of the technical terms used in it. Equally, you will gain a much better idea of the way that the author has approached the topic and how the experiments and data collected are used to support the thesis. It will also help you understand how they map on to previous interpretations and the broader context of the topic. Reading the complementary articles will also allow you to identify where there are contradictory results or interpretations. It is important that you identify where authors have conflicting views, and spell out the differences in your essay or project report. Where appropriate you should say which interpretation you prefer, and why it seems the best approach to you, using the appropriate source material to support your argument.

8. Memorize key authors' names

As far as any literature or other source material is concerned, the names of the authors and the titles are important. If the authors are those that are mentioned regularly, or identified in your reading list, then it is a fair bet that the book or paper is worth reading. When you are writing a project report or a final-year essay, you will probably have some idea of the standing of the authors and the reliability of their work. However, for a first-year essay it is always worth checking about the authors from your textbook or lecture notes. Sometimes, of course, an author is frequently cited because the results of the research are wrong or misleading. A well-known recent case is that in which a very limited study of twelve children suggested that the measles, mumps and rubella (MMR) vaccine might be involved in the development of autism (Wakefield *et al.*, 1998). A subsequent investigation by the General Medical Council (GMC) has shown that the results of the study did not match the hospitals' or general practitioners' records of the patients studied. In addition, major studies that followed have established that there is no link (NHS, 2008).

In a journal article the names of the authors are listed after the title. In some cases, particularly in those areas of physics that involve international research facilities, such as the Large Hadron Collider at CERN, the list of names of the authors may well run to seventy or more, and it is not uncommon in some areas of bioscience for the list to run to ten or more. In such cases your tutor or supervisor will know who the main author is. Normally the book or article will identify where the principal author is based. If it is a major university or institution, or a research group well known for the quality of its research, it will give you some feeling for the standing of the author.

9. Read most recent articles first and look at their quality

It is also useful to look at the date of publication of the book or article. If it is very recent, you should read it before others that were previously published to see how the author draws on earlier published research. With textbooks or research monographs you should also check that you are using the most recent edition. It is probably fair to say that where there have been many editions, the book is likely to be a standard source in the field and can be relied upon to provide the accepted approach. You will also find that certain journals have a higher reputation than others. Also research monographs published by a university press may be more highly regarded than those published elsewhere. Very often a research monograph will be published as part of a series and this is generally seen as a mark of quality. As far as books and research monographs are concerned, it may also be helpful to look at the reviews that were written when they were published. These can give you a good feel for the standing of the authors and the quality of the book.

10. Look at the titles

A book will always have a list of chapter headings, and very often a string of sub-headings for each chapter. A book will also have an index based on key words in the text. An index is particularly helpful in indicating where specific topics are mentioned in more than one chapter. Looking at the chapter headings and the index is always a good starting point before plunging into the text of the book.

The title of a journal paper should give you a fairly precise idea of what the research covers. Most titles will include specific key words. These are important, as experienced researchers will often start their background reading with an online search of these key words. A review article will usually have a table of contents, which in many ways will resemble the sub-headings of a book chapter. Reading the chapter headings and index of a book or the table of contents of a review article will help you to identify initially where the parts relevant to your essay or project are to be found, and so help you to focus your reading. Research papers, however, will not usually have a table of contents or even subheadings in the text. Where this happens, your starting point has to be the abstract, followed by a quick reading of the conclusion.

11. Read the preface or introductory chapter

When you begin to read the main text of a book, start by reading the preface. This will set out the structure of the book and will give some indication of the audience that it is aimed at, as well as saying why it has been written. Clearly, if it says that it is primarily aimed at graduate students, it will be more advanced than a general textbook, and is less likely to be appropriate as a starting point for a first-year essay. Sometimes, especially with research monographs, the preface will outline the contents of the chapters. Where it does not, you will generally find that the first chapter will form the introduction to the book. When you read the preface or introduction of a textbook or research monograph, you should also get into the habit of using the index to identify where the key words used in the preface or introduction crop up in the main text of the other chapters. Similarly with a journal article, you will start by reading the abstract. Very often this will include key words, and this will be especially helpful if the article is available online or published on a CD-ROM. The abstract of the article will also often be included in a published collection of abstracts, such as *Biological Abstracts*.

There are occasions when the title of the book or article does not reflect its true relevance to your essay or project, and reading the preface or introduction of the book, or abstract of the article, will help clarify this for you. It is also worth bearing in mind that the abstract part of a journal article may not summarize some important parts of the paper. An obvious case in point is where the abstract only picks out the key results of the research, and does not mention other experiments in the project that did not work. Information about these other experiments may be relevant to your essay or project. If you are in doubt about the relevance of the article, skim through the illustrations and data tables and the section describing the research methods used. These will generally show you much more clearly the relevance of the article.

It should be clear from the preface or introductory first chapter of a good book if it is relevant to your essay or project. They are obviously longer than the abstract in a journal article, and as such allow the author to be more expansive about the subject matter. A preface or introduction will also give a real feel for how readable the book is. If the author's style sends you to sleep, then you may want to think about whether it is sensible to continue, given that the following chapters may be equally unreadable or boring. In the preface or introductory chapter the main points should be clearly presented in a logical order. The language should be straightforward and easy to read.

12. Read on and take notes!

Once you have decided that the book or article is relevant, read on. It is important that you take notes that summarize the key points. Do not make long notes on sections that are not really relevant to your work. Start by summarizing the key points that the author is seeking to make in the chapter or article. These will usually be found in the opening paragraphs of the chapter, and will certainly have been identified in the preface or introduction to the book. In a journal article these will be found in the abstract and conclusion. Your summary of the paragraphs and sections of the main text will provide the supporting arguments for the author's proposition. If you find when you have reached the end of the article or chapter that there are contradictions, highlight them. Avoid copying down long quotations. There is always the danger that parts of your essay will end up as a string of unattributed passages, and this almost certainly will be seen as plagiarism.

Always summarize the key points in your own words after reading a whole paragraph or section. Summarizing a section or paragraph is the best way of learning to think about the material that you are reading. Inevitably there will be times when it will be difficult to summarize a passage, especially when, for example, in the description of the methods and materials used, the original source uses a lot of technical terms and is very dense. Nonetheless, do try to summarize it in your own words. This will help you to understand much of the technical language. If you come across any technical terms you do not understand, then note them down and look them up, for example, in the glossary or in a scientific dictionary. You can always seek the help of your tutor, though only when you have tried other sources. It is well known that we remember far more of what we find out for ourselves, than what other people have told us!

When you have summarized a paragraph or section re-read your summary to make sure that it reflects accurately the source material. Try to avoid paraphrasing, for example, by rewording a whole sentence in the original text. Avoid trying to take notes as you read, as you will be much more likely to copy out whole sentences or phrases. When you are taking notes, focus on the book chapter or journal article that you are reading. Do not be tempted to skip to another chapter or article that has been referred to in the text. Read them after you have finished reading your current text, otherwise you may end up with notes that confuse two or more sources. As far as journal articles are concerned, you will probably find it helpful to take a photocopy or a personal print-out, where it is an online journal, and use a highlighter pen when you have your first read through. Check the copyright rules for photocopying – your university library will have information on the regulations. A collection of hard copies with annotations and highlighted points will often be helpful when you are revising for your examinations, as well as for the essay or project in hand.

If you are used to working with a computer, whether a laptop or desktop, you may find it easier to key your notes in through the word-processor that you use routinely. There are advantages to this, in that you can subsequently use the word-processor to find key words in your notes. There is, however, the danger that where you are accessing an online journal, or a web-based article, you will simply cut and paste, and potentially end up with a large number of unattributed quotations, with the risk of plagiarism, when you come to write your essay or report.

13. Reading for your research project

The introduction of a journal article will normally set out the aims of the research conducted and the key elements of the background. This is particularly important as all research builds on previous work. The introduction should explain clearly why the author undertook the experiments, what research methodology was used, and how these relate to previous work. It should also set out the key results of the experiment, and how they fit in with previously reported work. The opening chapter of a research monograph will similarly give you a good idea as to how the research or matter described fits into the wider field of the topic you are studying. The introduction or opening chapter should always be read in conjunction with the bibliography, which will give you the references to previous work and the context for the present research.

When you are doing background research for a project or laboratory experiment, look at the techniques described in the article or book. In a journal article these will be found in the section about methods and materials. This will set out the research methodology, which will describe the techniques used as well as details of the materials used. This section may well be where technical language is used much more extensively, because each term has a precise meaning that other researchers will understand. Consult your supervisor as a last resort if you do not understand fully the terms used. Nonetheless, you need to recognize that by the time you reach your final year you will be expected to understand these terms fully. The methods and materials section will obviously give you another good starting point for your own experiments, and will allow you to evaluate the quality of data found in previous research. This may also help you to identify another and possibly more effective approach. It may also give some views on possible follow-up studies or experiments. This will again help when you are designing a project involving a range of experiments.

A word of warning: where an author has repeated a method already published elsewhere, they may not give full details, and may simply refer to the previous paper. For example the author may simply state 'the assay was performed according to Bloggs *et al.* (1999)'. This is frustrating because you may have to follow up the original reference to get the detailed method.

The discussion and conclusion sections of a research paper are clearly important because they set the research fully in context, and they will contain the detailed interpretation of the results. They will inevitably be the most subjective part of the text. You have to remember that the author is trying to convince the reader that it is an important piece of research, and that the experimental techniques and the approach used are appropriate and valid. Equally, the author is aiming to convince the reader that the integrity of the data stands up to scrutiny. This is also where the importance of the peer-review process comes into play. Look at the data set out in the paper, and compare them with the description of the results. In a well-written paper the author will always present the data or results obtained separately from the interpretation of them.

As has been said before, it is always important that other scientists are able to replicate the results, and where appropriate bring their own interpretation of them. Sometimes the researcher will have overlooked a possibly important interpretation, or have laid too much emphasis on a particular aspect, by misinterpreting the statistical implications, particularly where a sample in the study is small, with little or no control comparator. You should, therefore, look at the extent to which the conclusions are based on opinions rather than the evidence of all the data presented. This can be important where the author is trying to tailor the data to a preconceived opinion. A common fault is where the author presents only a very limited set of data that support the preconceived interpretation. When experiments are conducted, all the results should be available to other scientists. Increasingly material that has been obtained, but is difficult to fit into the main report, such as very detailed data, may be included in an appendix, or separately in an online supplement to the article. This is also the approach used when the experimenters have used videos to record material, for example, in a study of animal behaviour.

When you are looking at data presented in graphic form, make sure that you understand the statistical technique used in presenting the data. Most bioscience programmes include a module on statistics and data presentation. Look at your lecture notes and handouts on statistics. You should also look at the text related to the graphics or diagram to see how the researchers collected the data and what controls they used, especially where they are basing their claims on the statistical significance of the data presented. In a good paper the presentation and interpretation of the data will very often give you a good model for your own work. Badly presented or misinterpreted data can ruin a paper or report.

14. Recheck your summary

When you are reading an article as part of the background for an essay or project report, take particular care to ensure that your summary actually reflects what is written. Again, highlight in your notes any point that you feel is not supported by the data. Equally, highlight aspects where you think that there can be further follow up to the research described in the article. Above all, when you are writing your notes, remember that they will form the basis of your essay or report. It is clearly not sensible when you are actually writing the essay or report that you constantly have to go back to the original source material to confirm what you have written or to fill in gaps. Again, make sure that your notes include accurately the full title of the paper or book with page numbers where necessary, and the authors' names in line with the referencing conventions that you will be using. (See below for a more detailed description of the Harvard referencing system.)

Using material from the Internet

The Internet can be a helpful source, especially where the information comes from peer-reviewed journals available in electronic form. Indeed, in some disciplines the Internet may well be the first place where something is reported and the article may be the first version of a paper that will be published in hard copy later. Increasingly, too, online peer-reviewed journals are being developed. However, many of the sources on the Internet are not refereed, and may well be misleading. You should also be aware of the target audience of any article published on the Internet. For example, in biomedical or health science, an article about a disease is often aimed at the patient or the patient's relatives. Thus, the information, though correct, is not necessarily at the level required for your assignment. So, you should always be aware of **your** target audience, which is most often your tutor.

The last thing you should do is cut and paste indiscriminately from the Internet. For a start, this is easily detectable as plagiarism by the standard computer programs used by institutions. It is also possible that you may be inadvertently copying an unattributed passage from the work of your own tutor! It is always useful to seek the help and advice of your tutor or lecturer, and even fellow students, in sorting out your ideas, but in the end the words in the essay or presentation must be your own.

There are pitfalls in using the Internet as a source of information for your work. If you use Google or one of the other search engines, you will certainly be presented with a very large list of links. When the link is to a refereed academic journal, or an online peer-reviewed journal, you can be confident about the quality of the information (using Google Scholar will ensure that the links obtained are to reputable journals). Similarly, if the link is source material from a college or university, which will be identifiable by the use of. ac.uk or. edu in the URL, you can be reasonably confident about its quality. If it includes in the text or as a bibliography a range of references to original source material, then you can regard it as reasonably reliable. Nonetheless, you should always check it against the primary source material, such as peer-reviewed journals.

When it comes to other sources, you need to be more cautious. Wikipedia is an often used source and can be very useful as a starting point, but you have to remember that it is not formally peer reviewed, and there have been instances where rogue writers have added information that is downright inaccurate. Be aware that there are groups that will take a religious or political stance against certain material, and will try to alter it when the opportunity arises. Generally Wikipedia articles will include references to original source material, and wherever possible you should follow these up. However, these articles may be highly selective, and may not represent the whole picture. If you are ever in doubt, always go back to your textbook or lecture notes, and compare the references there with those in the online article. You should always take the same cautious approach to other articles on the web, even when they have a named author or authors. You will find from time to time that the text in these articles corresponds word for word with a similar article in another source like Wikipedia. In cases like this it is not generally possible to tell what the original text is and who is the real author. You may well find that the text looks like an unattributed paraphrase of another online source. Again, it is often difficult to identify the originality or the provenance of the text. As a matter of general principle, unless the online source is from a peer-reviewed journal, or from a clearly respectable academic source with solid references, be extremely cautious about using web-based material. Nonetheless, it can provide you with an alternative or unorthodox view about the topic you are researching, and as such you may wish to comment on it in your essay or report.

Plagiarism

Institutions and examiners in particular are constantly on the lookout for plagiarism. Plagiarism is deemed to take place when you fail to acknowledge properly the work of others you have used in your assignment. More specifically it is seen as plagiarism if you quote from or paraphrase an author's work without a reference, or where you have used the laboratory results of someone else without proper acknowledgement. It also includes the copying of other students' work, or work that is written by more than one student, when it is supposed to be a solo effort. Your own institution or department/school will have specific guidelines on plagiarism and the assessment regulations will detail the penalties if you are caught plagiarizing someone else's work. Plagiarism is taken very seriously nowadays particularly because of the ease with which students can download material from the Internet.

Referencing conventions

In your essay, presentation or report you will be expected to use consistently one of the standard referencing conventions to cite the sources that you have used in your background reading. You should therefore learn to cite and reference articles and books properly. In most conventions referencing takes the form of a two-stage process. The first is a citation in the text of your essay or report to indicate that you are drawing on information from the work of another author. The second is a reference, which gives more detail of the source to which you refer in your text. The detailed references to **all** the sources that you cite should be included in a list, usually placed at the end of your essay or report. In addition, there may be circumstances in which you will also include a bibliography in your essay or presentation. The bibliography will include a list of relevant sources that you have consulted in the process of preparing your paper or presentation, but not actually cited in the text. Very often the bibliography will be headed 'Further Reading'. This is a particularly useful way of referencing textbooks, review articles and abstracts. The format of the bibliography must always be the same as that of the list of references.

Different disciplines often have different styles of referencing (academics are like that: you can rarely get them to agree on a common convention across disciplines). For example, the conventions used in the biosciences will be different from those used in chemistry. In the UK many disciplines in the sciences and humanities, including biology, use what is known as the Harvard style. However, the Council of Science Editors (CSE) in the USA produces a substantial style manual that currently runs to 680 pages in its latest edition (Council of Science Editors, 2006). It recommends two systems of documentation: one that follows the Harvard style (also known as the name-year (N-Y) system); the other that follows the Vancouver style (also known as the citation-sequence (C-S) system). As usual, life is never simple: many of the journals in the biomedical and biochemistry areas follow the Vancouver style, for example, the *British Medical Journal* (BMJ). The conventions used by chemistry are those of the Royal Society of Chemistry, or the American Chemistry Society, which are broadly similar. Many university bioscience departments, however, recommend that students use the Harvard system, where references to the literature are cited in the text, the form of the citation is: (principal author(s), year of publication), for example (Smith, 1999). In the Vancouver style, citations are indicated by a number in parentheses in the text, for example, 'In a recent study (1)'. In Chemistry citations are indicated by superscript numbers in the text, for example, 'In a recent study[1]...'. In addition, each of the styles has a different convention for the format of the reference list or bibliography. Consequently in many areas of biology you are likely to draw on sources that use three different styles of referencing. You should always follow consistently the style recommended by your tutor. In this section we will focus on the conventions used in the Harvard style. A brief guide to the Vancouver system can be found on the British Medical Association's (BMA) website (British Medical Association, 2006).

The Harvard style of referencing

Up to the end of the nineteenth century there were no conventions for referring to the work or views of other authors in a book or article. For example, in his *Origin of Species* Darwin (1859) uses phrases like 'Professor Ramsay has given me ...'; 'from information given to me by Mr Watson ...' However, in the early 1880s Professor Edward Mark, who was Hersey Professor of Anatomy at Harvard University, and a very distinguished zoologist, used a name-year citation system, with a footnote to explain its purpose. The system was increasingly adopted in scientific articles. It seems, however, that the use of 'Harvard system' to describe it came much later and may have originated in the UK (see Chernin, 1988).

General principles

There are certain general principles that should be followed. Please note that the 'citation' refers to the detail given in the text, while the 'reference' refers to the item in the reference list at the end of the piece.

1. In the citation you should use only the author's surname without initials, unless you are citing another author with the same surname who has published an article in the same year as another.

2. The form of the citation will depend on whether you are emphasizing the name of the author, for example: 'King (1998) reported ...' or emphasizing the topic or research, for example: 'The aetiology of disease is ... (Ahmed *et al.*, 2007).

3. In the reference list the author's initials are used rather than the full first name(s). Do not give a mix of initials and full first name(s). The initial should come after the name. For example: Watson, J.D. and Crick, F.H.C.

4. In the citation give both names if there are two authors. For example: 'Watson and Crick (1953) proposed a structure for DNA'. If there are three or more authors, cite the first author and use *et al.* (which is a shortened Latin form for 'and others').

5. When a government department/agency or other public organization is referred to, the organization's name is generally used in the citation, unless there is a named author.

6. Where a book is part of a series, the series title and series number should be given after the book title in the reference.

7. Where more than one place of publication is listed (for example, Cambridge, London, New York), include only the first named in the reference.

8. The reference list should be in alphabetical order, based on the name of the first author, or name of organization.

9. With multiple-authored works, always list the names in the reference in the order in which they appear in the source material.

10. Where a book has been reprinted, only enter the first published date as the year of publication. However, if a new edition has been published, reference the latest edition, unless you are actually using an earlier edition.

11. Where an article has no date, put 'no date' where you would normally cite or reference the year. This is relatively unusual in printed works, but may occur on websites where an organization is referring to an earlier publication.

12. You need to distinguish between chapters in an edited book, and chapters in a multi-authored book. In an edited book, the citation will be author of the chapter and date, and the reference will make it explicit that it is a chapter in an edited book, as well as giving the name of the editor(s). In a multi-authored book the citation will take the form '(Jones *et al.*, 2003)' or 'Jones *et al.* (2003)' and the reference will give the full list of authors.

13. Where the author has not been identified, as is generally the case with government White Papers, the accepted practice is to give the name of the organization where you would give the name of the author. If the name of the chair of the report committee is usually added to the title (for example, 'the Leitch Report'), cite the organization and date.

14. For electronic copies of books or journals on disk, it is normal to reference them in the same way as the printed versions. However, it is increasingly the case that primary sources, particularly journals, are

being made available through the Internet. Where the Internet version has been consulted, it is now quite normal to reference the URL.

15. Where material is available on the Internet it is the established practice to give the full URL, rather than the general URL for the publisher or organization followed by an instruction 'and follow the link to ...'. However, this can be a problem when the material is on a password protected site. In addition you should also give the date that you accessed the source in the reference. There are slightly different conventions for the order of the date of accession and the URL. Some institutions use the form 'Available from: URL ...' followed by '[date of access]'; others use the form '[date of access], URL'. You should always follow one convention consistently throughout your reference list or bibliography. This will normally be the convention recommended by your institution, but remember, if your report or presentation is going to be published, then you should follow the convention of the publisher concerned.

16. Where you refer to papers in journal articles or given at conferences that have been published online, the normal practice is to reference them in the same way as the printed version, but with '[online]' inserted after the tile of the journal or conference. In addition you should also give the date that you accessed the source in the reference and the URL in line with one of the conventions listed above.

17. Where a PDF version of a paper or report is available, in the reference this should be given explicitly as the URL. However, sometimes the URL of the PDF version is not given. In such cases reference the URL of the page that you accessed, and indicate 'available as a PDF' at the end of the reference. You must remember that the page numbering of a PDF file source is likely to be different from that of an HTML or other file source, so it is important that you reference accurately the source that you used.

Quotations and how to cite and reference them

When you are using a direct quotation from an author or organization, you should always cite the author's or organization's name in the form: name, year of publication: page number (for example, Ahmed *et al.*, 2007:203). If you are quoting an online source, substitute 'online' for the page number. The quotation should always illustrate a point that you wish to make, and should always use the exact words of the original. If you quote only part of a sentence use three

dots (…) for the part that you have omitted. If you want to add a couple of words
to the quotation, for example, to help clarify the quotation, put the addition in
square brackets (for example, 'The Director [Professor John Smith] indicated
…'). Only add to the quotation where it is necessary for clarification.

Where a quotation is less than a line of text, enclose it in single quotation
marks. Where you give a quotation within a quotation, always use double quotation marks to identify the internal quotation.

If the quotation extends to more than a single line of text, begin a new paragraph and indent it. The part of the sentence that you have written before the
quotation should end with a colon. The actual quotation in the indented paragraph should not have quotation marks.

When you do quote directly, as well as citing the author in the text you should
always give a full reference in your reference list or bibliography.

Using software to generate your references

There are software programs, like Endnote, and word-processing programs,
such as Microsoft Word 2007, that will allow you to create references in different styles. The range of styles that Word 2007 will allow you to use is limited,
and does not include the Harvard system. It does, however, include the American
Psychological Association (APA) style, that is very similar. Endnote includes
a much wider range of options. There are, of course, advantages in using a
word-processor that has an embedded citation and referencing system. Word
2007 has a specific tab labelled 'References'.

Examples of citations and references

The principal formats for citations and reference lists (including bibliographies)
are set out in Tables 2.1–2.3. There are formats for printed works, such as books
and journal articles, and for electronic and multimedia material, though you
should remember that references to electronic versions of books and journal
articles will follow the conventions used for print versions. Examples are given
to indicate how you should use the conventions. The list of examples is not
exhaustive: there are, for example, ways of referencing blogs, email sources or
personal interviews. In some disciplines these will be used both as primary and
secondary sources. There are useful and easily accessible online guides that
will give you details of how to cite and reference these other sources (Leeds
Metropolitan University, 2009; Patel *et al.*, 2009; Anglia Ruskin University,
2008; Imperial College Library, 2008).

Table 2.1 How to reference: books, including textbooks, research monographs and edited volumes

Type of source referred to in text	Citation format	Citation example	Reference format	Reference example
Work with a single author or two authors	Surname(s) (year) *or* (surname(s), year)	As King (1998) reports … It has been reported that … (King, 1988)	Surname(s), initial(s) (year of publication) *Book title.* Series title and number, edition [*where applicable*]. Place of publication: publisher.	King, D.J. (1998) *Applications and Engineering of Monoclonal Antibodies.* London: Taylor & Francis.
		As Price and Newman (2001) report … From a recent experiment … (Price and Newman, 2001)		Price, C.P. and Newman, D.J. (2001) *Principles and Practice of Immunoassay.* London: Macmillan.
Work with three or more authors	Surname of first named author *et al.* (year) *or* (Surname of first named author *et al.*, year)	Ahmed *et al.* (2007) show that … The aetiology of disease is … (Ahmed *et al.*, 2007)	All authors' surnames, initial(s) (year of publication) *Book title.* Series title and number, edition [*where applicable*]. Place of publication: publisher.	Ahmed, N., Dawson, M., Smith, C. and Wood, E. (2007) *Biology of Disease.* Abingdon: Taylor & Francis.
Two authors with the same surname; different publications	Initial, surname (year) *[for each]* *or* (initial, surname, year [*for each*]) *[Shown in chronological order]*	J. Smith (2005) and A. Smith (2008) have both … There is evidence that … (J. Smith, 2005, A. Smith, 2008)	*[As for single-authored work* **but listed in alphabetical order, not chronological order]**	

Table 2.1 (*Continued*)

Type of source referred to in text	Citation format	Citation example	Reference format	Reference example
More than one publication of a single author in different years	Surname (year, year) *or* (surname, year, year) *[Shown In chronological order]*	The findings of Smith (1999, 2006) … There is evidence that … (Smith, 1999, 2008)	*[As for single-authored work **but listed in chronological order within surname**]*	
In the same year	Surname (yeara, yearb) *or* Surname (yeara) and Surname (yearb) *[when cited on different pages or in different paragraphs]* *or* (surname, yeara, yearb) *[the a, b, c, suffixes are assigned in the order that they appear in the essay]*	The findings of Smith (2006a, 2006b) … Smith (2006a) reported … *[cited in paragraph 1 or page 1 of your essay]* Smith (2006b) has shown … *[cited in paragraph 2 or page 12 of your essay]* There is evidence that … (Smith 2006a, 2006b)	*[As for single-authored work **but listed by the a, b, c, date suffix that they appear in the essay or report and not in chronological order**, for example, a book published in April may be listed before one published in January]*	
More than one source referring to similar techniques or outcomes	Surname (year) and surname (year) *or* (surname, year; surname, year) *[in alphabetic order]*	Chapel (1999) states … This view is backed by Goldsby et al. (2003) and Todd et al. (2005) It is generally accepted that … (Chapel, 1999; Goldsby et al., 2003; Todd et al., 2005)	*[As for single-authored/ multiple-authored work **and incorporated in alphabetical order in the reference list**]*	

More than one source, but including more than one paper by the same author	Surname (year, year), surname (year) and surname (year), *or* Surname (year), surname (year) and surname (year) *or* (surname, year, year; surname, year; surname, year) *[In alphabetic order but with the works of the same author listed chronologically. If the works of the same author are in the same year, use the a, b, c, year suffix as appropriate]*	The findings of Allen (2001, 2003), Smith *et al.* (2001) and Williams and Jones (2004) all suggest … The findings of Allen (2001) suggest that … Other research conducted later by Allen (2003), supported by the work of Smith *et al.* (2001) and Williams and Jones (2004) suggest that … There is general agreement among the main researchers (Allen, 2001, 2003; Smith *et al.*, 2001; Williams and Jones, 2004) …	*[As for single-authored/ multiple-authored work and incorporated in alphabetical order, in the reference list, but with the papers of the same author listed in chronological order within surname]*
Where an author is cited as a single author in one publication and as the first author in a second one	Surname (year) *[for the single-authored publication]*, surname *et al.* (year) or surnames (year) for the multi-authored publication *or* (surname, year *[single-author publication]*; surnames, year *[multi-authored publication]*	The findings of Allen (2001) suggest that …, while other research conducted later by Allen *et al.* (2003) … There is general agreement (Allen, 2001; Allen *et al.*, 2003)	*[As for single-authored/ multi-authored work but with the single-authored publication listed before the multi-authored one]*

Table 2.1 (*Continued*)

Type of source referred to in text	Citation format	Citation example	Reference format	Reference example
Where another author is cited by the source you have read, but you have not read the cited source *[If you have read both the main article and the cited source cite both in the normal way.]*	Surname (year, cited in surname, year) *or* (surname, year, cited in surname, year)	The findings of Allen (2001, cited in Jones and Williams, 2003) suggest that … There is general agreement (Allen, 2001, cited in Jones and Williams, 2003) …	*[List the main source as for single- authored/ multi-authored work, in this case the Jones and Williams' publication]*	
Chapter in an edited book, but the author is not one of the editors	Surname(s) of the chapter author(s) (year) *or* (surname(s), year) of the chapter author(s)	Dawson and Moore (1989) identified … A major issue (Dawson and Moore, 1989) is … *[Do not list it as if it were a secondary citation, or simply by citing the editor's surname]*	Surname(s), initial(s) (year of publication) Chapter title. In initial(s) and surname(s) of editor(s), *Book title,* series title and number, edition [*where applicable*] (chapter pages). Place of publication: Publisher.	Dawson, M.M. and Moore, M. (1989) Immunity to tumours. In I. Roitt, J. Brostoff and D. Male (eds), *Immunology,* 2nd edn (pp. 1–18). London: Gower.

52

Table 2.2 How to reference: other printed sources

Type of source referred to in text	Citation format	Citation example	Reference format	Reference example
Journal articles where the names of the authors are given	Surname(s) (year) *or* (surname(s), year) *[In multi-authored papers, the standard Smith et al. format is followed]*	Etzioni (2003) emphasizes … There is clear evidence (Etzioni, 2003) that … Ismail and Snowden (1999) set out very clearly … It has been established that … (Ismail and Snowden, 1999)	Surname(s), initial(s) (year of publication) Article title. *Journal name,* **volume number** (issue number) *[as available]*, page numbers of article. *[Very often the titles of the journals are abbreviated in line with established conventions]*	Etzioni, A. (2003) Immune deficiency and auto-immunity. *Autoimmun. Rev.,* **2**, 364–369. Ismail, A.A. and Snowden, N. (1999) Autoantibodies and specific serum proteins in the diagnosis of rheumatological disorders. *Ann. Clin. Biochem.,* **36**, 565–578.
Journal articles where the name of the author is not given *[for example, editorial comment or review, or in popular journals]*	(Author unknown, year)	In a recent article in *The Railway Magazine* (author unknown, 2009)	Publication name (year of publication) Article Title. *Journal name,* **volume number** (issue number) *[as available]*, page numbers of article.	*The Railway Magazine* (2009) £1bn electrification plan for Scotland. *The Railway Magazine,* **155**(1302), 6.
Newspapers (named author)	Author(s) (year) *or* (Author(s), year)	Boseley (2009) in a recent report in the press … As recently reported in the press (Boseley, 2009) …	Surname(s), initial(s) (year of publication) Article title. *Newspaper name,* supplement title *[where applicable]*, date of publication, page numbers of article.	Boseley, S. (2009) Doctor to go before medical council over claims of 'ghost writing' for US drug company. *The Guardian,* 19th September, p.8.
(unknown author)	(Author unknown, year)	A recent press report (author unknown, 2009)	*Newspaper name* (year of publication) Article title. *Newspaper name,* supplement title *[where applicable]*, date of publication, page numbers of article.	*Manchester Evening News* (2009) Fears over 'deadly' herbal remedy. *Manchester Evening News,* 6th October, main edition, p.30.

Table 2.2 *(Continued)*

Type of source referred to in text	Citation format	Citation example	Reference format	Reference example
Paper delivered at conference (published)	Surname of author (year of publication) *or* (surname of presenter, year of publication)	In a report of research delivered recently, Dawson and Overfield (2005) … The conference report included details of a recent study (Dawson *et al.*, 2009) …	Surname(s), initial(s) (year of publication) Conference paper title. In name of organisation. Conference title [*including annual conference number, if stated*], volume number [*where applicable*], Location of conference [*including venue and city*], date of conference. Editor's (or chair's) surname, initial [*where available*]. Place of publication: publisher, page number(s) of article.	Dawson, M.M. and Overfield, J.A (2005) Plagiarism: do students know what it is? In Proceedings of the Science Learning and Teaching Conference 2005, University of Warwick. Published jointly by the Higher Education Academy Subject Centres for Bioscience, Materials and Physical Sciences, pp. 166–167.
Paper delivered at conference (not published)	Surname of author (year of conference) *or* (surname of presenter, year of conference)	In a paper detailing research delivered recently at a conference, Dawson *et al.* (2009) … A recent conference report included details of a study (Dawson *et al.*, 2009) …	Surname(s), initial(s) (year of conference) Conference paper title. Paper presented at Conference title. Location of conference [*including venue and city*], date of conference.	Dawson, M.M., Forsyth, R. and Ready, R. (2009) *Developing cross-institutional communities of academic practice – from Groundhog Day to The Wizard of Oz.* Paper presented at HEA Annual Conference, Manchester, 30th June, 2009.

Source	In-text citation format	In-text example	Reference list format	Reference list example
Guest presentation	Surname of presenter (year of presentation) *or* (surname of presenter, year of presentation)	In a presentation of research delivered recently, Dawson and Overfield (2008) … A recent presentation lecture included details of a new study (Dawson and Overfield, 2008)	Surname(s), initial(s) (year of presentation) Lecture title. Paper presented at Conference title. Name of institution where the presentation took place, city, date of presentation.	Dawson, M.M. and Overfield, J.A. (2008) Teaching students what plagiarism is. Paper presented at the HEA Centre for Bioscience event 'Preventing and Designing out Plagiarism', University of Leicester, 8th April 2008.
Government papers and other official sources both in the UK and elsewhere (such as WHO and UNESCO)	Department/Organization name in full [*on first use*], followed by abbreviation and year of publication. Sometimes the report will generally be known by the name of the enquiry's chair.	In a recently published review, Leitch (2006) …	Government Department/Organization (year of publication) *Title.* Cmnd. Official reference number [*where available*]. Place of publication: publisher (name of chair, and 'Report' or 'Review' [*as appropriate*].	HM Treasury (2006) *Review of Skills: Prosperity for all in the Global Economy – World-Class Skills.* London: HMSO (Leitch Review).
Government department notes or information papers	Department/Organization name in full [*on first use*], followed by abbreviation and year of publication.	In a note published by the Parliamentary Office of Science and Technology (POST) (2002) …	Government department/organization (year of publication) Article title. Series title [*where appropriate*], Volume number/Issue number) [*where applicable*], page number(s) of article. Place of publication: publisher.	Parliamentary Office of Science and Technology (POST) (2002) Peer review. Postnote, No. 182. London: POST.
Thesis/Dissertation	Author's surname (year of submission)	Recent research undertaken by Dawson (1976) …	Surname(s), initial(s) (year of submission) Thesis title. Degree submitted for. Name of awarding institution.	Dawson, M.M. (1976) *Humoral Immunity in Human Neoplasms.* PhD. University of Manchester.

Table 2.3 How to reference: electronic and multimedia sources

Type of source referred to in text	Citation format	Citation example	Reference format	Reference example
Electronic books on disk	*As for printed books*	*As for printed books*	*As for printed books*	*As for printed books*
Consulted through the web	Surname(s) (year) *or* (surname(s), year)	The advice of Barass (1984) … Students are generally advised that … (Barass, 1984)	Surname(s), initial(s) (year of publication) *Title. Series title and number, edition [where applicable]* [e-book]. Place of publication: publisher. Available from: URL [date of access]. *or* [date of access] URL.	Barass, R. (1984) *Study! A Guide to Effective Study, Revision and Examination Techniques* [e-book]. London: Chapman & Hall. Available from: http:// books.google.co.uk/books?id= FIg9AAAAIAAJ&dq=Barrass+ Study!+a+guide&printsec=frontc over&source=bl&ots=IZhYnyL Mn4&sig=jZ4rqt607tsellyhNAH WMuUVRHw&hl=en&ei=pkO3 SuGnNIef4gammYx9&sa=X& oi=book_result&ct=result&resnu m=3#v=onepage&q=&f=false [Accessed September 2009].
Electronic journals on disk	*As for printed journals*	*As for printed journals*	*As for printed journals*	*As for printed journals*
Consulted through the web	Surname(s) (year) *or* (surname(s), year)	Chernin (1988) gives a brief history … The history of the Harvard style … (Chernin, 1988)	Surname(s), initial(s) (year of publication) Article title. *Journal name* [online], **volume number** (issue number) [*as available*], first and last pages. Available from: URL [date of access]. *or* [date of access] URL.	Chernin, E. (1988) The 'Harvard system': a mystery dispelled. *BMJ* [online], **297**, 1062–1063. Available from: http://www. pubmedcentral.nih.gov/ picrender.fcgi?artid=1834803& blobtype=pdf [Accessed September 2009]. Also available from: http://www. bmj.com/cgi/pdf_ extract/297/6655/1062 [Accessed October 2009].

Newspapers	As for printed version	Surname(s), initial(s) (year of publication) Article title. *Newspaper name* [online], Supplement name [*where applicable*], publication date, page numbers of article. Available from: URL [date of access]. *or* [date of access] URL.	Corrigan (2009) in a recent report in the press … In a report of academic plagiarism (Corrigan, 2009) …	Corrigan, C. (2009) Is academic plagiarism being hidden?*The Guardian* [online], Education Guardian, 28th July, p.3. Available from: http://www.guardian.co.uk/education/2009/jul/28/academic-plagiarism [Accessed August 2009].
Articles from websites (where the author is named)	Surname(s) (year) *or* (surname(s), year)	Surname(s), initial(s) (year of publication) *Title* [online]. Available from: URL [date of access]. *or* [date of access] URL.	Loy (2002) gives a brief account of the dispute between Newton and Leibniz … The dispute between Newton and Leibniz (Loy, 2002) …	Loy, J. (2002) *Newton vs. Leibnitz* [online]. Available from: http://www.jimloy.com/calc/newtleib.htm [Accessed August 2009].
(where the author is not named)	Organization (year) *or* (organization, Year)	Organization (year of publication) *Title* [online]. Available from: URL [date of access]. *or* [date of access] URL.	The NHS (2008) has dismissed claims that autism is linked to the MMR vaccine … Claims that autism is linked to the MMR vaccine have been dismissed (NHS, 2008) …	NHS (2008) *MMR vaccine 'does not cause autism'* [online]. Available from: http://www.nhs.uk/news/2007/january08/pages/mmrvaccinedoesnotcauseautism.aspx [Accessed August 2009].

Table 2.3 (*Continued*)

Type of source referred to in text	Citation format	Citation example	Reference format	Reference example
Databases	Organization that has the database (year) *or* (Organization that has the database, year)	Data from the European Bioinformatics Institute (part of the European Molecular Biology Laboratory) (EMBL-EBI) (2009), newly released … Data from EnsemblPlants, a recently launched a web resource (EMBL-EBI, 2009), …	Organization (year of publication) Database title [online]. Available from: URL [date of access] *or* [date of access] URL.	The European Bioinformatics Institute (EMBL-EBI) (2009) EnsemblPlants [online]. Available from: http://plants. ensembl.org/index.html.
DVDs/ Videocassettes	Title (*abbreviated if necessary*) (year of distribution)	Firefly Luciferase degradation slowed by inhibitor compound (see illustration: Ades and Auld, 2009)	Title of DVD/Videocassette (year of distribution). Directed by [DVD] or [Videocassette]. Place of distribution: distributing company. *[If it has been saved from a TV programme give the name of the TV channel (for example, BBC4) and date of broadcast]*	*Testament of Youth*. (1996) Directed by Armstrong, M. [Videocassette]. London: Acorn Video.
Online images or photographs	Surname/ organization (year of publication)	Firefly Luciferase degradation slowed by inhibitor compound (see illustration: Ades and Auld, 2009)	Surname(s), initial(s) (year of publication) *Image title or description* [online image]. Available from: URL [date of access]. *or* [date of access] URL.	Ades, J. and Auld, D. (date unknown) *Firefly Luciferase Degradation Slowed by Inhibitor Compound* [Online Image]. Available from: http://www. genome.gov/Images/press_photos/highres/20149-300.jpg [Accessed October 2009].

58

References

Anglia Ruskin University Library (2008) *Guide to the Harvard Style of Referencing* [online]. Available from: http://libweb.anglia.ac.uk/referencing/files/Harvard_referencing.pdf [Accessed October 2009].

Boseley, S. (2009) Doctor to go before medical council over claims of 'ghost writing' for US drug company. *The Guardian*, 19th September, p.8.

British Medical Association (BMA) (2006) *Reference Styles: Harvard and Vancouver* [online]. Available from: http://www.bma.org.uk/library_medline/electronic_resources/factsheets/LIBReferenceStyles.jsp [Accessed September 2009].

Chernin, E. (1988) The 'Harvard system': a mystery dispelled. *BMJ*, **297**, 1062–1063 [online]. Available from: http://www.pubmedcentral.nih.gov/picrender.fcgi?artid=1834803&blobtype=pdf [Accessed September 2009]. Also available from: http://www.bmj.com/cgi/pdf_extract/297/6655/1062 [Accessed October 2009].

Council of Science Editors (2006) *Scientific Style and Format: The CSE Manual for Authors, Editors, and Publishers*, 7th edn. Reston, VA: CSE in conjunction with Rockefeller University Press.

Darwin, C. (1859) *On the Origin of Species by Means of Natural Selection*. London: John Murray.

Fleischmann, M. and Pons, S. (1989) Electrochemically induced nuclear fusion of deuterium. *Journal of Electroanalytical Chemistry*, **261**(2A), 301–308.

Imperial College Library (2008) *Citing and Referencing Guide: Harvard Style* [online]. Available from: http://www.imperial.ac.uk/Library/pdf/Harvard_referencing.pdf [Accessed August 2009].

Leeds Metropolitan University, Skills for Learning (2009) *Quote, Unquote: A Guide to Harvard Referencing* [online]. Available from: http://www.library.mmu.ac.uk/eres_targets/didsbury_harvard.pdf [Accessed August 2009].

NHS (2008) MMR vaccine 'does not cause autism' [online]. Available from: http://www.nhs.uk/news/2007/january08/pages/mmrvaccinedoesnotcauseautism.aspx [Accessed August 2009].

Patel, H., Shields, E., Padma, I. and Beck, N. (2009) *Harvard Referencing* [online]. Available from: http://www.library.mmu.ac.uk/eres_targets/didsbury_harvard.pdf [Accessed September 2009].

Wakefield, A.J., Murch, S.H., Anthony, A., Linnell, J., Casson, D.M., Malik, M., Berelowitz, M., Dhillon, A.P., Thomson, M.A., Harvey, P., Valentine, A., Davies, S.E. and Walker-Smith, J.A. (1998) Ileal-lymphoid-nodular hyperplasia, non-specific colitis, and pervasive developmental disorder in children. *The Lancet*, **351**(9103), 637–641 [online]. Available from: http://briandeer.com/mmr/lancet-paper.htm [Accessed October 2009].

Watson, J.D. and Crick, F.H.C. (1953) A structure for Deoxyribose Nucleic Acid. *Nature*, **171**, 737–738.

Further reading

Bioscience Horizons, a national e-journal for the publication of undergraduate research. Available from: http://biohorizons.oxfordjournals.org/.

Carpi, A., Egger, A.E. and Kuldell, N.H. (2008) *Scientific Communication: Understanding Scientific Journals and Articles*. Available from: http://www.visionlearning.com/library/module_viewer.php?mid=158 [Accessed September 2009].

Egger, A.E. and Carpi, A. (2009) *Scientific Communication: Utilizing the Scientific Literature* [online]. Available from: http://www.visionlearning.com/library/module_viewer.php?mid=173 [Accessed September 2009].

Olin and Uris Libraries, Cornell University (2009) *Critically Analyzing Information Sources*. Available from: http://www.library.cornell.edu/olinuris/ref/research/skill26.htm [Accessed September 2009].

Parliamentary Office of Science and technology (POST) (2002) Peer Review. *PostNote*, no. 182. London: POST.

The Vision Learning Website, which is supported by the US Department of Education, has a number of very useful articles. Available from: http://www.visionlearning.com/).

3

Essay Writing

About this chapter

In this chapter we look at how essays are structured, and the importance of planning. We give some hints on researching your essay (linking back to Chapter 2) and stress the need to answer the question that was set. We suggest some general rules to follow, and things to avoid. We also give a few tips on writing essays under examination conditions and for achieving first-class answers.

The purpose of assignments

Your university or college department will set a variety of assignments such as essays, reports and poster presentations during the three or four (or more) years of your programme. Some of these assignments will be *formative*, that is they are not assessed formally, while others are *summative* and the marks will count towards your final grade in the unit or module in question. All assignments are important, whether the marks count or not. This is because assignments have other purposes in addition to grading. You may be given a particular assignment in order to develop your skills, for example, in searching the literature or perhaps in written communication. Ideally, the process of completing an assignment will aid your learning, not hinder it. Your tutor should provide details about:

- whether the assignment is summative or formative;
- what skills are being tested;

Communication Skills for Biosciences   Maureen M. Dawson, Brian A. Dawson and Joyce A. Overfield
Copyright © 2010, John Wiley & Sons Ltd.

- what learning outcomes are being tested in this assignment;
- practical information, such as the word length of an essay, or the size of a report or poster;
- the assignment deadline and penalties to be applied if it is handed in late.

Why write an essay?

The formal essay is an integral part of most undergraduate programmes in universities and colleges in the UK. Essays are often required at various stages in degree and other programmes, and the aims and objectives will reflect the stage at which the essay has to be written. A third-year (level 6) essay will generally be tougher than a first-year (level 4) one. When your department or school sets and marks an essay, your tutors will assess it at the appropriate level. So, an essay that gets excellent marks when submitted in your first year will not get you good marks if it is submitted in your final year. During your university course you will progress, and more is expected of you at each stage. On your side, when you are writing an essay, you will learn to research the range of available background material relating to a specific topic, and to study, review and assess independently the literature you have found relating to the topic of the essay. As a result of writing the essay you will acquire generic communication and presentational skills and improve on them as you progress through your course. Reading the literature will give you some feeling of the language, style and conventions of your discipline.

It is highly likely that you will have to write an essay at an early stage in your programme. Although you will probably have had some experience of writing essays before starting your current course, you may find that what is required is generally both quantitatively and qualitatively different from what you have done earlier. Most essays will have a maximum word length which is generally larger than you may have come across before. Your tutor will also expect you to go beyond the basic textbook material and to draw on the scientific literature of your discipline. Writing an essay will generally involve you in individual work, either in the library, or working online, and, while you may get some help from your tutor or supervisor about source material, you will be expected to develop your own techniques for reviewing the relevant literature. At school, too, you may have been used to handing in successive drafts of your work, with comments from your teacher about how you could improve this piece of work. In university, commenting on draft essays is unusual, and you are unlikely to get a second chance to improve it. This does not mean, however, that the comments you get back are not useful, as these comments can be taken into account when you submit your next essay or assignment.

Universities and colleges recognize the importance of essay writing as part of the education process, and will give all kinds of guidance about what is required, ranging from formal requirements for length and layout to much more informal tutorial guidance about source materials and so on. They also stress the importance of giving formal and informal feedback (comments on your work) to students. If you feel you need more feedback on your work, you should ask your tutor.

Skills

Essay writing involves a range of generic skills and techniques and many of the techniques that you learn can be applied across any discipline or in your later career. In your first year, the research that you do for your essay will introduce you to using your university library as an information resource. This will stand you in good stead for all your assignments in the future. You may well have gained some experience from using Internet search engines to find information on specific topics, for example, for your A-levels. However, the range of resources available through a university library will generally allow you to develop much more evaluative search skills. In other words, you will learn to distinguish good academic sources from the wealth of other sources available that may be inappropriate for all sorts of reasons. University libraries have access to more scientific literature online than is available through the general Internet. Essay writing will teach you to evaluate material and information and, where appropriate, to criticize information obtained from the literature. In areas of the biosciences where knowledge and facts can lead to different interpretations, you may be required to write an essay which sums up and evaluates different theories and opposing viewpoints. A good example of this might be to support, with evidence, a particular theory of evolution. In this case, you will need to have a coherent and logical approach to ensure that your interpretation can be supported by the facts. Essays in scientific disciplines are largely on topics that have involved the research of a great many academics over the years. However, researchers often disagree on interpretations of the evidence and they, too, need to provide evidence to support their case.

Preparation

In this section we outline the stages you should go through when preparing to write your essay and probably the most important of all is the first stage!

1. Start in plenty of time

Most tutors will tell you the submission date of an essay well in advance. This submission date is just that – it is not the date when you start to write your essay and neither is the night before the submission date. You will need to prepare well in advance, so that the essay you submit is the best you can do.

2. Read and understand the title

It may sound obvious, but make sure you read the title of the essay and are confident that you understand what it is about. If you have any doubts, talk to the tutor who set the assignment. By all means ask another person on your course, but they may not necessarily give you the right answer. Always read any guidance notes provided. When you have done this, you will be in a position to begin the real process of preparation for the specific topic of your essay.

3. Read around and be critical about your sources

Your tutor may have suggested some literature and, if so, a good start is always a recommended up-to-date textbook. This will provide you with a broad outline of the topic. It will also have a useful bibliography, which can form the basis of your literature search. Once you have identified and read initial key papers, you can then look at other sources on the basis of the bibliography in those papers. Most papers will cross-reference to twenty or thirty other relevant papers and you will find after reading two or three that many will cross-reference the same background sources. In this way you will get a very broad-based introduction to the topic, as well as the means of identifying the critical papers relating to it. It is generally said that for a good final-year essay a student will have read up to thirty articles, including review papers and original research papers. It is also useful to remember that in many disciplines there are published abstracts, such as *Biological Abstracts*. You would not normally use them for a first-year essay, but later on, particularly when you begin project work, they are invaluable.

You should aim to get information from the literature, and bring it together from a range of sources. You need to use the information that you have gained as the basis for the discussion of the particular topic or problem, or range of problems, and to set out your ideas clearly and simply to the person who will read your essay. As you will find, essay writing brings together both the learning process and the development of independent thought and interpretation. Scientists generally use published papers to set out and communicate their ideas to others. Essay writing will help you to develop the ability to communicate ideas and concepts to other people. You will find that, as with many skills, practice makes perfect.

Given that the preparation will include having to read a number of papers/ articles, it is always sensible as a starting point to read the summary and conclusion of each article/paper. In this way you will avoid the detailed reading of irrelevant articles needlessly. A clear structure and a rigorous approach to reading will also help you to avoid muddling or misinterpreting information, which would inevitably lead to the 'wrong' answer. You will often find that so-called scientific articles in newspapers are wrong or misinformed, largely as a result of sloppy reading of the original scientific paper.

It is always very useful to get a feel for the literature in your discipline. An article that is easy to read and understand will provide a good stylistic model for your work. If it seems dense and packed with technical terms, then it may well be badly written, or is aimed specifically for an expert audience in a limited field. Where you find an author's style comfortable, try to pick out what you like about it. As often as not, you will see that is based on the use of clear and uncomplicated sentences with balanced phrases.

Wherever you can, you should base your essay on original research papers, particularly in your final year. Chapters in books, and particularly textbooks, are really very long review articles, where the authors give their own interpretation of the original source material. Research papers in academic journals describe original research carried out by the authors. These papers will have been peer-reviewed before publication. What this means is that other researchers have read the article and 'refereed' the work to ensure that it is not flawed. This is seen very much as the gold standard for reporting research.

Use a textbook as a starting point, but remember it is not the final authoritative source. Make sure that the textbook is up to date. A book published this year may have been three to four years or more in development. Again, articles in popular science magazines or in encyclopaedias should be handled very carefully, as should material in online encyclopaedias, such as Wikipedia. Newspaper and broadcast media reports are sometimes misleading, inaccurate or skewed. Often the press will quote word for word (*verbatim*) and without attribution from press releases issued by university or research institute press offices. It is not easy to spot such instances. A report may begin with a sentence like 'Scientists at a Manchester university have made a major breakthrough in combating cancer'. The text that follows may well be a *verbatim* quote from the press release, or, as often as not, an exaggerated summary based on a misinterpretation by the journalist of the statistical data released.

Always be discriminating about your sources. Remember that drug companies often employ ghost writers to write papers to put results of drug trials in the most favourable light. Papers ghosted in such a way can have a disproportionate impact on meta-analyses, where data from a range of papers in a related area are analysed to identify common factors and outcomes. Sometimes such published data will be included in a large, computer-based metadata analysis, and may skew the outcome significantly.

4. Make notes from the literature

Note taking is an important part of preparation. It allows you to develop your skills at summarizing the key points of an article or chapter. You should include in your summary material that is immediately relevant to your essay and that which may turn out to be relevant at a later stage. Aim to make your notes after reading a whole paragraph or even the whole article, rather than taking notes as you read – that way you will avoid copying word for word instead of summarizing. When you are looking at a bibliography or references, always copy them down accurately. As much as anything this will help you to learn the referencing conventions, and will save time later, when it may be more difficult to find details for an incorrectly copied or incomplete reference.

When you are reading a textbook or paper, you should always think about what you are reading. You may find that in some cases the author generalizes, but fails to give specific examples that support the arguments or ideas being set out. Again, keep an eye out for inconsistencies – especially in textbooks or review articles – and, 'where appropriate', note down your concerns.

5. Make sure you know how an essay is structured

A scientific essay has the following structure:

- an **introduction** setting out the topic and your approach to it or a summary of your line of argument, together with a brief note of the historical background to set the context. Your essay title may have terms which need defining – this is the place to do it. Depending on the word length of the essay, an introduction is often a single paragraph in which you 'set the scene';

- a more **detailed description** of the subject, often involving tables of data and illustrations. This is the 'meat' of the essay. You may need to set out your argument and the evidence to support it. It is important that this part is logically constructed and is always relevant to the title of the essay. This is where you will be expected to cross-refer to the literature;

- a **conclusion**, generally including a summary of your more detailed description, and bringing together different strands. In some cases it will include a summary of issues that have not yet been solved;

- a **bibliography** or **list of references** that you have consulted as part of the preparation for your essay. It is important that you include references to all the material that you have consulted, especially as unattributed material included in the detailed description or conclusion may be regarded as plagiarism. Make sure you know whether you are required to produce a bibliography or a reference list. Often, in the first year, students are asked to list the books they have used to prepare the essay – this is the bibliography. A reference list is more precise – you will need to cite the reference in the text and in the reference list and there are several ways of doing this (see Chapter 2). In-text referencing is required, certainly for second and final years, but often also in the first year.

6. Make a plan based on the outline structure and the major points gained from your reading

You will need to plan your essay at an early stage and develop it as you read more background information. It helps enormously if you have a broad outline and structure of what you want to cover in your essay. A useful tip is to start with a set of bullet points, which you can adapt and expand as you read more about the topic. These bullet points will eventually form the headings and subheadings of your essay. Alternatively you may wish to produce a mind map. A mind map or concept map is a graphical method of taking notes. Having a visual representation of what you want to include can help organize your essay. Mind maps generally have a tree branch format, with branching successively from a central idea. The visual aspect of mind maps can help to gather ideas and structure your essay prior to setting it within a formal essay structure. Mind-mapping programmes are available; though these are not essential, they do make 'playing around' with your ideas much easier. A very simple (and incomplete) example for an essay entitled 'DNA' is shown in Figure 3.1. This has been prepared using a programme called Inspiration 8 IE (Inspiration Software Inc).

Now, take another look at the essay title. As well as indicating the topic of the essay, it will always contain key words that indicate the style of the essay. For example, is it a simple essay where you are asked simply to *describe* a structure or a process, or are you being asked to *discuss* – in which case you will have a lot of scope for your own interpretation? An essay title which contains the words *the development of* ... will be a historic review, and as such will be very different from, for example, *current approaches to* ... An essay which requires you to *compare and contrast*, for example, chloroplasts and mitochondria, needs a lot of thought. It should not be half an essay on the chloroplast and half on the mitochondrion.

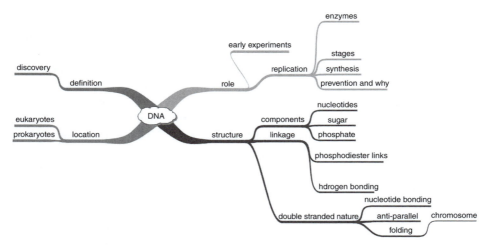

Figure 3.1   A mind map based on DNA

Set out your ideas within the essay structure. It may help if you do this on a computer, because, having set things out, you may feel you need to swap things around. Writing a good plan is time well spent, because once you have done that, writing the essay will be a lot easier.

It is important that you bear in mind your audience: in most cases you will be writing for someone who knows about the topic. However, there will be times when you will be asked to write with the general reader in mind. Consequently, in your introduction, and in the text more generally, you should define any technical terms. Technical terms are essentially a shorthand way of expressing a complex concept within a discipline. By defining them you will also show that you have understood the key elements of the topic of your essay. As you will have found in preparing your essay, a list of references provides vital background material for the readers, which will allow them to follow up and check particular details. The techniques that you learn in writing your essay are vital practice if you wish to become a professional scientist. The techniques for referencing have been discussed in Chapter 2. You should get guidance from your department/school about which particular referencing convention to follow.

7. Write your essay based on your plan

You will probably be asked to word-process your essay, especially as you may be required to submit the assignment electronically to the virtual learning environment (VLE) that your university is using. Your word-processor has lots of features that can help you structure your essay, particularly in relation to normalizing fonts for text and headings, and text style generally. Knowing about these features will save a lot of time when formatting. Think also about other aspects of your structure, such as your style, paragraphs, sentences and avoidance of plagiarism. As you write, you should also bear in mind the essay title, and check after each paragraph that what you have written is relevant to that title – it is very easy to go off the point and, particularly in a word-limited essay, you need to make each word count.

The paragraph

Many people have problems with paragraphing and any such problems will detract from your essay. The paragraph is always the basic unit of an essay. A paragraph should contain points that are interlinked or relate to a particular element of the wider topic. It is always worth stating early in the paragraph what the theme of it is. If you are setting out a particular point, you should include in the paragraph the relevant facts, interpretation and conclusion in a logical order. However, when you are bringing in a new concept, then you should begin a new paragraph. Equally, the sequence of paragraphs should be organized logically, so that the reader is able to follow your thinking. It is particularly helpful to have phrases that link together sentences and paragraphs, such as: 'initial thinking about ...'; 'by contrast ...'; or 'in the light of the data it is now clear...'.

Always think about your reader. If, for example, you summarize two different approaches to a problem without flagging up the differences, even at a simple level, then there is the danger that you may confuse your reader. It is helpful to have pointers such as: 'Pauling asserted ...'; 'by contrast Darwin believed ...'. Equally, you should give an indication of where your own view lies, particularly where you are giving an evaluation of different theories. Where you are making a statement or argument, make sure that it is backed by the facts, or by the evidence. Your analysis is an important part of your essay, and shows that you have been thinking about the topic. Each paragraph should build into a logical subsection, and so on upwards, so that you eventually move to a conclusion where each part is seen to develop from its predecessor. When you are jotting down your bullet points as part of your initial preparation, you should look out for those that can be grouped together and can build one from another.

A paragraph should normally be about half a page in length. Do not, as some students do, change paragraphs after every sentence.

The sentence

The key component of the paragraph is the sentence. To return to basics: a sentence starts with a capital letter, contains a verb and ends with a full stop. However, as Sir Ernest Gower (1954) points out in his book *The Complete*

Plain Words, 'A sentence is not easy to define'. Sir Ernest wrote his book with the aim of moving the civil service away from the use of the sort of fancy language that was commonly known as mandarin. Official reports nowadays generally follow his principles, and consequently can be read easily. If the civil service can do it, so can you!

Use short sentences wherever you can. Some authors write long complex sentences where the exact meaning can be difficult to gauge. Indeed, it used to be the case that authors routinely used fairly high-flown language and often repetitious phrases in their sentences. This is no longer so prevalent, and short, simple sentences will help you to get your meaning across. Try to use active verbs rather than passive forms and avoid using obscure and fancy words. For example: instead of writing 'The eradication of X is often brought about by Y', simply write 'Y often eradicates X'.

One exercise you might like to try is to take an editorial leader from a 'broadsheet' newspaper such as *The Times* and translate it into simple English. You may find that the editor dresses up commonplace points in fancy language in an attempt to impress the reader. Using simple language is especially important in an essay, if you wish to convey your points clearly. Try reading your final draft aloud. If it sounds pompous and high-flown to you, or if you find yourself getting lost in a particular sentence, then you should simplify your text. Even better, read your essay out loud to a friend; if they find it difficult to understand, then you are failing to communicate clearly.

When writing a sentence, make sure that subject and verb, as well as indicative adjectives and nouns agree in your sentences. For example: 'The number of examples is …' or 'These are the causes of …'. Always watch out for words that are Latin or Greek plurals: for example: 'These phenomena are …' not 'This phenomena is …'. Another common example is the use of the word 'data': 'data' is a plural word, the singular form of which is 'datum'. You should say 'data are …' or 'the data show'.

Try to build sentences that are based on a main clause and a subordinate clause, rather than two or more clauses joined by 'and' or 'but'. For example, rather than writing 'The cat sat on the mat and licked its paws' you might write 'As the cat sat on the mat, it licked its paws'. Organize your sentences carefully. Where you are following on from information in a previous sentence, start your new sentence with any linking material, and put your new information at the end. This will help ensure the logical sequencing of your paragraph. For example:

> During his time as a naturalist on *HMS Beagle*, Darwin noticed significant variations within the same species as he travelled from island to island in the Galapagos. It was this variability that led him towards his theories on the origin of species.

Stylistically it also helps if you use parallel and balanced phrases in your sentences, rather like bullet points, but written as continuous text.

There is a general view that the end of a sentence is where you make a point that you want to emphasize, though this is not always the case. For example:

> Because of his work on the origin of species, Darwin is generally regarded as one of the leading thinkers of his generation.

is equally valid as

> Darwin is widely regarded as one of the greatest thinkers of his generation because of his book *The Origin of Species*.

It really does depend whether you want to emphasize how Darwin is regarded, or what made Darwin famous.

General points of style

1. Whenever you can, use a simple word rather than a pretentious synonym; for example: 'It started to rain', rather than 'the precipitation commenced'. Inevitably, as a bioscientist you will have to use technical terms. As often as not, technical terms are based on Latin- or Greek-derived words, and have very specific meanings, which convey complex concepts in a single word or phrase. Wherever possible when you use a technical term for the first time in your essay, you should say in simple words what it means. Your reader may well know the term, but by spelling out its meaning in simple terms, you will show your reader that you understand the underlying concept.

2. Where you use an abbreviation you should follow the following conventions:

   – with some abbreviations you should use a full stop to indicate that it has been abbreviated. An example of this would be in the name of a species, if you have already given the name in full. So, for example, '*Staphylococcus aureus*' or '*Escherichia coli*' become '*S. aureus*' and '*E.coli*'. Another good example occurs in reference lists, when the first names of authors are abbreviated: 'Dawson, M.M., Dawson, B.A. and Overfield, J.A.', for example. A further example is when using the shortened forms of Latin terms: '*e.g.*', '*i.e.*', and '*etc.*', are further examples, though in an essay it is always better to write these out in full;

– where the abbreviation is an upper case acronym, you should not use a full stop to indicate an abbreviation. An acronym is a pronounceable name for an entity such as a group or organization, made from the initial letter of the words making up the group. An example is *UNESCO* which stands for the United Nations Educational, Scientific and Cultural Organization. Generally you should write the name of the body in full when you first use it, followed by the abbreviated form in brackets; for example, 'Engineering and Physical Sciences Research Council (EPSRC)', particularly where it is a body that may not be immediately recognized by your reader. After that you can simply use the acronym;

– units of measurement are not followed by a full stop.

3. In a word-processed essay, italics are often used for foreign words, for example, the Latin words used for genus and species. There are recognized conventions about generic and specific names. They must be italicized, and the name of the genus must begin with a capital letter, for example: '*Oenanthe oenanthe*' (Wheatear). If you are giving the name of another species in the same genus, you can abbreviate the genus, thus for example: '*O. hispanica*' (Black-eared Wheatear). Another example is the use of '*in vitro*' and '*in vivo*'. If you are handwriting your essay you should underline the relevant words.

  Titles of books or papers are also normally italicized (or underlined).

4. When using numbers you should generally use words, rather than numerals in your text; for example: 'sixty samples were taken'. The exception is where you want to give a measurement, or date, when you will normally use numerals; for example: '500 mg' or 'September 1st, 1990'. When you wish to refer to a percentage, use words in your text; for example 'sixty percent'; but use numerals and the percent symbol elsewhere, such as in tables or illustrations. Be consistent in the way in which you use numbers/numerals throughout your essay, and where appropriate follow the conventions of your discipline.

5. Tidiness is important. For essays written as coursework you should use a word-processor, and follow any guidance on layout, font, page size etc. given by your department/school. For essays written under examination conditions, take care that your handwriting is clear and legible. If you do have any specific disability that causes you difficulties in essay writing, you should inform your department/school at the earliest opportunity and get appropriate support.

6. Depending on the word limit, you may wish to use sections and sub-sections and this can be helpful in guiding the reader. This will rarely

be needed in a 1500-word essay, although it might be used in a 3000-word essay. If you have any doubts, ask your tutor – academic staff often have differing views on this practice!

7. If you use tables and figures, then they should be numbered separately. A photograph, graph, drawing or diagram counts as a figure. All figures and tables should be referred to in the text, before they appear. The figure or table should have a title which goes above or below the actual figure. For example: 'Figure 1: A diagram to show the nitrogen cycle'.

   A word of caution: it is very easy nowadays to obtain figures from the Internet. However, simply cutting and pasting a figure may not be appropriate for the following reasons:

   – a figure that looks good on a computer screen very often appears pixellated when printed;

   – figures obtained from Internet sources may be overcomplicated and may not help your description;

   – figures obtained from Internet sources often have inappropriate legends and the figure numbers will not match yours.

   You should always give the source of the figure in the figure legend, to avoid accusations of plagiarism.

8. It always helps to have at least one draft before you write your final version. (Obviously you would not do this for an essay written under examination conditions!) Never leave essay writing to the last minute. It helps enormously to re-read a draft after a gap of a few days. It is also very useful to have someone else look at your essay before you finalize it. At a minimum they will pick up spelling and grammatical errors. If you are using a word-processor, use the UK English spell-checker. If in doubt use a dictionary, particularly where there are two words that sound the same, for example, 'peddle' and 'pedal'. Proof-read your final draft, particularly checking for punctuation errors. Number your pages: it will help your tutor if the pages get mixed up after you have handed them in. Some tutors are now using audio record-ings to give feedback on your work and need to be able to refer to page numbers. Always keep a copy of your final text. If you are using a word-processor, it is also very helpful to keep a note of the version of your draft; including the date in the file name is a useful way to do this.

9. In many disciplines there is a convention that you do not quote directly. If you are writing an essay on, say, the language of the Greek playwright Aristophanes, direct quotation is essential. However, the normal convention in the science disciplines is that you should set out the general concepts in your own words, unless, of course, the original author has used a phrase or passage that is the standard accepted description. In such cases you should use your own words to make the essential point, and then back them with the relevant quotation. If you do quote directly, make sure that the quotation is accurate. When you are quoting, you should always make sure that you attribute the quotation to the author and reference it (see the notes on plagia-rism below). Another word of caution here: an essay that entirely consists of a series of quotations from textbooks is not plagiarism if the quotations are attributed. However, this is bad practice and will get you poor marks – your reader wants to hear your 'voice' not those of a dozen other authors.

10. When you are writing your essay, try to avoid the use of clichés, and unsupported or inaccurate generalizations. For example, if you are writing an essay on influenza, it is not sensible to write sentences like: 'Swine flu is a killer disease, which spreads like the plague'. Instead you should be more expansive and support your thinking with data or facts along the following lines:

> Influenza variant H1N1, generally referred to, even in official
> publications, as 'swine flu', is seen by many members of the

public as a killer disease. Statistically the latest available data show that. … These death rates are significantly lower than those for seasonal influenza. Inevitably comparisons are made with the great influenza pandemic of 1919, in which up to 21 million people are believed to have died. (Global data for this pandemic are generally seen as unreliable, because in many countries the cause of death may well not have been properly recorded.) In addition, the views of many people will have been coloured by recollections of the influenza epidemics of the 1950s. Clearly a major factor in the initial transmission of the disease from country to country is the relative ease of international air travel, though a number of countries, like the US and China, attempted to isolate travellers who showed symptoms of the disease and kept them under close observation. Most influenza viruses are spread by person-to-person contact, in contrast with epidemics of other diseases, where transmission is often by other intermediate hosts, such as rat fleas or mosquitoes. A vaccine against H1N1 has been developed, and countries like the US and the UK plan to use it in mass vaccination programmes as a means of limiting the spread of the disease. However, the World Health Organization (WHO) has identified a number of under-developed countries, which do not have the cash resources to pay for the amount of vaccine required for mass vaccination, or have only a limited health service infrastructure. Consequently the impact of a local epidemic could be significant: local infection rates and death rates may well approach or even pass those of the 1919 pandemic, with major economic and social consequences, particularly where health services are already overburdened by other diseases like malaria or HIV/Aids.

11. Finally, always follow the standard conventions used in your department or school for citing source material.

Plagiarism

Until relatively recently plagiarism was seen as stealing the ideas or research of another academic. Accusations of stealing concepts and information are not uncommon in the history of science. One of the most notorious disputes involved Isaac Newton and the German philosopher Gottfried Leibnitz, who developed quite independently the underlying principles of calculus, though in different forms. Newton called his version 'the method of fluxions and fluents'. Although Newton discovered the principles before Leibnitz, he did not publish his ideas until later. There is also clear evidence that the two corresponded regularly over the period when their ideas were being defined, and Newton described his methods in letters he wrote to Leibnitz. If anything, these probably helped Leibnitz to build on his own initial and independent thinking. In the bitter wrangling that followed in the early part of the eighteenth century, Leibnitz appealed to the Royal Society. However, Newton was President of the Royal Society and appointed the investigating committee, which is said to have consisted entirely of his friends. He himself wrote the committee's report, which accused Leibnitz of plagiarism. The report was formally published by the Royal Society as *Commercium Epistolicum* in 1713. To compound his intrigues, Newton was subsequently the anonymous author of a review of the committee's report, which was later published by the Royal Society. One of the consequences of the dispute was that Leibnitz's approach was largely ignored in England for most of the eighteenth century. Ironically Newton was never consistent in his notation. By contrast, Leibnitz developed a meticulous method of notation that is generally used nowadays. (Professor Stephen Hawking has a very nice account of Newton's obsessive behaviour in his book *A Brief History of Time*.)

Nowadays, plagiarism is generally seen as a form of cheating, in which material is copied without attribution to the original author. There is no doubt that the easy availability of material on the Internet that can be cut and pasted into documents has led to a significant increase in the copying of text and data without attribution. However, you should remember that by citing your sources, you are actually providing evidence that you have read widely around your topic. In addition, as has been said before, your citations will help the reader of your essay to follow up and check details.

Definitions of plagiarism in college or university regulations generally include guidelines to help students avoid plagiarism in their work. The following guidelines are essentially a summary of good practice based on those of a number of universities, including the University of Leeds, the University of Manchester and Monash University. In summary, you should not use quoted material, either single phrases or longer passages, without indicating that it is a quotation, and without indicating the source of the quotation. Where you paraphrase or summarize material, again, you must always provide a reference for the source of your material. You should never copy the work of fellow students (see also the note about collusion below). It is never wise to purchase and copy an essay from any of the online providers of model answers (generally known as essay banks). Apart from the fact that it is dishonest to pass someone else's work off as your own, even if you have paid for it, you will not acquire any of the skills that doing coursework provides. In addition, your tutor will discover the cheating when using plagiarism detection software.

If you do use data or illustrations from someone else's experiments in your school or department you should always acknowledge it, even if it has not formally been published or reported. Your work should not be jointly written with others, unless it is specifically a collaborative project for which a joint report is expected. In joint reports it is important that you identify your own contribution to the work as well as that of others, and to ensure that it is clear to all that it is a collaborative effort. Always ask for guidance from your tutor or supervisor about the way you should present joint work. In addition you should not submit work that you have already submitted for an assignment for a second assessment, perhaps in another unit or module. This is regarded as self-plagiarism.

Your university will have its own very specific guidelines on plagiarism, which will be available from your department/school. In particular, these will emphasize that it is important that any essay, paper or presentation is the student's own work (or students' in the case of joint projects). These conventions will not stop you from quoting or summarizing material from the scientific literature. (If you think about it, this is an integral part of a written essay.) What you must do is acknowledge the source of your material by way of a formal reference. Plagiarism is currently seen as a major problem by many universities and many routinely use software to detect plagiarism in students' essays and coursework.

Related to plagiarism is a form of cheating generally referred to as 'collusion', where you use someone else's work, with their permission, but without attribution, as part of an essay or project. An obvious example is where you copy from an essay that a friend has submitted in a previous year, or for a different programme of study, for which he or she obtained a high mark. Equally if you let someone else copy a piece of your work, especially when you got a high mark, both you and the other person are likely to face disciplinary procedures for cheating. Always bear in mind that any form of cheating can lead to your getting a zero mark for your essay. If you are found to have cheated in a piece of work or a project that contributes to a degree or diploma assessment, your overall classification could be downgraded. In exceptional circumstances your degree or diploma might even be taken away from you.

Examination essays

As the Boy Scout's motto says 'Be Prepared'!

The hints and tips for writing essays in examinations assume that you have prepared fully for the exam by doing plenty of revision. You should revise the whole syllabus, not just material that you 'think' will come up in the exam.

Question spotting is a high-risk strategy that can lead to failure, especially if those few topics that you have revised fail to turn up. If you have revised the whole syllabus, you will be able to make links across the subject which demonstrates your ability to draw connections.

Reading what is actually written on the examination paper is absolutely vital. Always read the requirements set out at the top of the examination paper. This is called the 'rubric'. It will inform you about the number of questions to be answered, and the time allowed for the examination. Allow a couple of minutes to read the rubric. From this information you can calculate the time you should spend in answering each question. You should also factor in time to structure and review each question you answer. For example, if you have two hours to answer four questions, allow five minutes to structure each answer before you begin the essay, and twenty minutes for the actual writing of the essay. Set aside the balance of the time that you have available (in this example this amounts to about twenty minutes) to review what you have written at the end of answering the four questions. You may need to modify your calculation of the time available for each answer to take account of any weighting that the examiner may give to a particular question. For example, there may be a compulsory first question with a weighting of forty percent of the marks for the paper, with a further three optional questions each with a weighting of twenty percent. In these circumstances you should allow more time for the compulsory question. When you have reached the end of the time that you have set aside for each question, you should finish your sentence or paragraph, even if you have not covered all the points that you want to make, and leave a gap in the answer book. You should then start your answer to the next question and aim to complete the remaining questions within your timetable. Normally, the examiner will give more marks for four answers, even if one or two are incomplete, than for three complete answers. At the end you can, of course, use part of the time that you have set aside for reviewing what you have written to add more to your incomplete answer(s).

Once you have identified the overall time frame for each answer, read through the detailed questions. If there are particular questions that you think you understand and can answer reasonably well, concentrate on them. Underline the key words in each question that you intend to answer. Try putting the question in your own words, and then look again at the wording of the original question. If they mean the same, then you have at least understood what the examiner requires. It is surprising how many students fail to read a question properly, and consequently give poor answers.

As with coursework essays look out for words like *discuss, compare and contrast*, or *outline current approaches to*. These will tell you how you should structure your essay. Some will require an answer that is a straightforward rehearsal of the facts (for example, *outline* …). Others will require you to take a critical approach to the question (for example, *discuss* … or *compare and contrast* …).

Many departments give students the opportunity to practise examination questions, and previous examination papers are generally available online. Take every opportunity to look at them, together with any model answer that your department or school may provide, and practise writing answers to the questions in your own words. However, it is not wise to learn model answers by heart at the expense of revising the subject, as questions do vary, and your skill will be to answer the question set, not the one that you have learned.

When you have selected the questions you believe you can answer (together with any compulsory questions), make a bullet point summary of your answer for each question. It is a good idea to practise this when you are revising. When you re-read your lecture notes and textbooks and so on, get into the habit of summarizing the key elements as bullet points. It can be helpful to draft your summaries for all the questions you aim to answer at the start, combining all the five minutes that you have set aside as preparation time for each question. If you have, say, twenty minutes available cumulatively, spend fifteen of them drafting your initial outlines. Then, as you come to answer each question, look at the set of bullet points for that essay, and add to or modify it as necessary before you start writing. If you have overrun your schedule on writing your answers for the first three of your four essays, and only have, say, ten minutes available for your final essay, focus on tidying up, expanding and ordering the bullet points that you originally drafted, and set them out, with a brief introduction, as your answer. In this way you will at least have an outline of your answer that the examiner can mark. Equally, if you only have a very limited time at the end to review your answers, and some are incomplete, use expanded bullet points to complete them.

As with coursework essays you should structure each essay so that it has an introduction, a main discussion section and a conclusion. You will not, of course, be in a position to provide a bibliography. Your introduction should be a single paragraph that sets out the key points of your essay. The main discussion section should amplify these key points in a logical order. Each point should be supported by detailed information and references to original papers and authors where you are able to provide them. If you have prepared thoroughly for your examination, you will be able to name the key authors and papers for the topics covered in the examination paper. Finally, your concluding paragraph should draw together the key points, and emphasize the view that you want to put forward in your essay. In a sense, it is a restatement of the introductory paragraph, but coloured by the detailed material that you have introduced in your middle section and emphasizing your views on the topic.

It is not uncommon to come across a topic in an examination paper that you have covered in a term-time essay, or laboratory report, or in your review of past examination papers. If this happens, be especially careful when you read the question: it is quite likely that the current question is a tweaked version of the earlier one. You will, of course, have the advantaged of being well informed about the detailed background of the topic, but you will have to tailor this to the new form of the question. You should also be careful not to go overboard on a question that you know a lot of information about. Remember to be selective, and keep within your time limit for writing your answer.

In your written examination essay you will have to be more concise and focused than you have been when writing a term-time essay or laboratory report. It is very easy to fall into the trap of padding out your answer with superfluous information. If anything, this may indicate to your examiner that you have not really digested and understood the question. The information that you give should always be focused on the key points you identify in your introduction. The opening sentence of your introductory paragraph should be based on the wording of the question, and is a statement of your thinking about the topic. Each of the following paragraphs that make up your middle section and in which you amplify the key points of your introductory paragraph, should begin with a sentence that sets out simply the key point. The remainder of the paragraph should contain all the supporting material for this point. You should not split the material that supports a key point identified in an introductory sentence over more than one paragraph.

If you stick closely to the timetable that you have set for writing your answers, you should have time to go back to complete any partial answers. Nonetheless, you must ensure that you still have time to review all your answers. As part of this review, check your punctuation, and look out for misspellings, garbled phrases or incomplete sentences. Do not attempt to rewrite any answer.

Above all remain calm and do not panic at any stage in your examination. Your examiner will know from personal experience that examinations can be stressful, and consequentially the marking of a written examination essay is rather less strict than that of a term-time essay or laboratory report, though any factual inaccuracies will not be ignored.

Further reading

Buzan, T. (2000) *The Mind Map Book*. London: Penguin Books.
Carroll, J. (2008) Credit where credit is due: citation and plagiarism. In G. Hall (ed.), *A Handbook for Postgraduate Research Students*. London: Palgrave Macmillan.

English Department, University of Birmingham (2002) *How to Write an Essay* [online]. Available from: http://www.english.bham.ac.uk/staff/tom/teaching/howto/essay.htm [Accessed August 2009].

Gower, E. (1954/1972) *The Complete Plain Words*. London: Pelican/Penguin.

Hawking, S. (1988) *A Brief History of Time*. London: Bantam Press.

On Line Writing Lab (OWL), Purdue University [online]. Available from: http://owl.english.purdue.edu [Accessed August 2009].

School of Archaeology, University of York (2007) *Guide to Writing Essays and Other Projects* [online]. Available from: http://www.york.ac.uk/depts/arch/ugrad/bos/essays.htm [Accessed August 2009].

School of Biological Sciences, Monash University (2005) *Essay Writing: Guidelines for the School of Biological Sciences* [online]. Available from: http://www.biolsci.monash.edu.au/undergrad/docs/essay-writing-guidelines.doc [Accessed August 2009].

University of Leeds (no date) *Plagiarism – University of Leeds Guide* [online]. Available from: http://www.lts.leeds.ac.uk/plagiarism/what_is_it.php [Accessed August 2009].

University of Manchester (2008) *Guidance to students on plagiarism and other forms of academic malpractice* [online]. Available from http://www.campus.manchester.ac.uk/medialibrary/tlao/plagiarism-guidance-for-students.pdf [Accessed August 2009].

4

Writing Practical Reports

About this chapter

In this chapter we discuss how to write up a formal practical report. We also mention some other forms of laboratory report that may be required in a work situation. The practical session may have taken place within a laboratory, but it could also have been undertaken in the field, as part of an ecology exercise, perhaps during a field trip. No matter where the session has taken place, the report will be more or less the same.

Introduction

In Chapter 1 we saw that scientific method is about:

- making observations;
- constructing hypotheses;
- carrying out experiments to test the hypotheses;
- analysing the data from the experiments;
- making further hypotheses.

Thus, all the scientific knowledge we have today has been built up by the work of our scientific predecessors who have communicated the results of their

Communication Skills for Biosciences   Maureen M. Dawson, Brian A. Dawson and Joyce A. Overfield
Copyright © 2010, John Wiley & Sons Ltd.

experiments by writing them up. Writing a practical report is therefore one of the most important things a scientist does because of the following:

- what they did;
- why they did it;
- what happened;
- what they found out;
- how this has added to knowledge of the system;
- how they could have done it better;
- what they will do next.

In addition, by allowing other scientists to read the report (peer review; see Chapters 1 and 2) the scientist has the opportunity to get their interpretations of the results, which may differ from his or her own, and to get further suggestions for future work. Also, if the methodology has been correctly reported in sufficient detail, others will be able to carry out the experiment to confirm or refute the result.

For you, personally, writing a practical report will give you time to reflect on your experiment and to see how your results relate to the original hypothesis. Your report will enable your tutor to learn what you did and what you learned in the process. Moreover, if you are carrying out an extended piece of research, which might be a three-week mini-project in your first year, or an extended year-long project in your final year, writing up each experiment thoroughly will enable you to keep a record of your work which you can revisit in the future. Do not imagine for one moment that you will always be able to remember why you did what you did six months ago, unless you have it all written down in your laboratory log book. Similarly, you will no doubt lose those scraps of paper that you wrote your results on – so make sure you transfer the results to your laboratory book!

In this chapter we will cover the basic components of the practical report and the style of writing that is used. Although we use the term 'laboratory report', remember that experiments are often undertaken outside the laboratory.

Some general points

Although we have discussed scientific method, you may not always feel that this is what you are doing in the laboratory. This is because your practical work will have different purposes. Early on, probably in your first year, your laboratory work may revolve around learning and/or practising new skills. The exercises you are given enable you to learn by demonstrating known facts. The accuracy and precision of your work may be tested, for example, by asking you to determine the concentration of an analyte[1] in a solution, or the number of cells in a suspension. Once these skills have been demonstrated, you may be lucky enough to undertake one or more mini-projects – these will have some research base and will probably be undertaken in small groups.

To summarize, your laboratory work will have the following purposes:

- to learn and practise new skills;
- to demonstrate known facts and relate to the theoretical aspects;
- to find out new things (the most exciting!).

As you go through your course, the amount of real 'experimentation' will increase, so that, by your final year, you will probably be doing an individual research project.

What to do before the practical session

Do not leave it until you write up the practical report to find out what you should have done in the practical session. You will probably have had access to the practical schedule before the actual class. The best way to get the most out of your practical session is to have read the schedule, to make sure you know what you have to do. The schedule itself will have an aim, introduction and detailed methodology. Your tutor may also have given you some references relating to the work you are going to carry out. If so, make sure you read them before the session – not just when you are writing up. It is also a good idea to check the assessment details for the session you are going to undertake as this may have a bearing on how you conduct the practical exercise. For example, the tutor may ask you to submit your report at the end of the session, in which case you need to build in time for writing.

If you are working with a partner, then it is also a good idea for you to go through the schedule with your partner – you can learn from each other. Check also whether the assessment involves submitting an individual or a joint report.

---

[1]An analyte is a substance being determined in an analytical procedure.

Make sure you take any equipment with you that has been specified – these will most likely include a laboratory coat and a pair of protective spectacles. Many laboratories will send you away if you do not have these vital pieces of safety equipment, and all your preparation will have been in vain. It is always useful to have with you a semi-permanent marker pen, which can be used to label glassware. Make sure you bring a notebook and pen, to take notes and to record your observations.

During the practical session

Your tutor will probably wish to give the whole group a brief introduction to the work being undertaken. He or she will also want to point out any changes to the practical schedule (make sure you write these down) and details of hazards and safety procedures to be followed. If these are not already in the schedule then make sure you write them down. Your tutor may refer you to any COSHH forms that you need to read. COSHH stands for 'Control of Substances

Hazardous to Health'. It is part of UK criminal law that chemicals and procedures involving hazardous chemicals must have a risk assessment, which not
only assesses the risk but also details how to reduce the risk both to yourself
and to other workers. In your first year, in standard laboratory experiments,
these assessments will have been carried out for you and your duty is to read
the COSHH form so that you know how to carry out the procedures safely.
Later on in your course, and especially when undertaking final-year projects,
you will have to carry out these assessments yourself, and complete the form,
so it is as well to find out about them early on.

If you have already read the schedule and associated references, the introductory session will reinforce your knowledge of what you have to do. If you have
not prepared for the session, then you may be confused, and this may very well
lead to mistakes in procedure.

If working with a partner, ensure you plan the work carefully so that both of
you are working towards the result and that you both get a chance to practise
any new skills. One person doing all the work and the other writing down the
results, or, worse still, simply watching, does not make for a good division of
labour, neither will the onlooker learn or develop his or her own practical skills.
You should record any ways in which you have varied from the schedule as
these may explain any unexpected results. Make sure you both have a copy of
the results before you leave the laboratory. Alternatively, if one person has taken
all the results, establish a procedure for ensuring that the partner will get a copy
– before you leave the lab.

How should it be written?

The way in which the laboratory report is written is fairly standard throughout
the bioscience disciplines. The writing should be clear and concise; do not use
three words where one will do. You should aim to tell the reader what you were
setting out to do, and give sufficient detail to enable the experiments to be
repeated. The structure of the report will be as shown in Figure 4.1. In addition,
some guidelines for completing each section are given below.

Please note that there may be minor variations from this structure. For example,
for an individual practical the 'Title' may take the place of the 'Aims'. The Aims
section is also sometimes given after, rather than before, the 'Introduction'.
Although these variations make very little difference to the overall structure
of the report, it is best to follow any guidelines given to you by your tutor.

Be careful about the tenses that you use as they may vary between sections
but never within a section. This will become clearer when you read about the
individual sections. Be careful not to plagiarise (see Chapters 2 and 3). Make
sure that any material you have sourced from elsewhere is cited in the text and

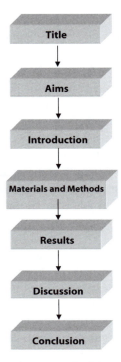

Figure 4.1   Structure of a laboratory report

fully referenced in the reference list at the end. The sections where you are most likely to use source material are the 'Introduction', 'Materials and Methods', and the 'Discussion'.

A word of warning ...

You may well have worked with a partner in the laboratory to carry out the experiment. However, unless you have been told that you can submit a joint report (with both your names on it), you should write up the reports separately. It is very easy for markers to spot where students have colluded to prepare a report (for example, the same mistakes will have been made, and very often the same phrases used). You may be accused of collusion, and the mark may be withheld. You must also never allow someone to copy your work and submit it as his or her own (equally, never copy another student's work). The person who copies is guilty of plagiarism, and, unless he or she owns up to it, you are both in danger of being penalized. The temptation to copy arises often because the student has left the assignment too late and has no time to check his or her understanding of the report.

Title

This simply describes, very concisely, what you are going to do. You should keep it short – leave the detail for the aims, for example:

'Determination of the activation energy of an enzyme'

'Investigation of factors affecting the activity of an enzyme'

Aims

This section tells the reader, usually in a single sentence, what you are setting out to do. As already stated, this may differ slightly from the Title. For example, if the Title was 'Investigation of the factors affecting the activity of an enzyme', the Aims might be more precise: 'To study the effect of temperature and pH on enzyme activity'. However, for single, one-off practicals, the Title and Aims are often interchangeable.

Introduction

This section provides the background for what you are going to do in the practical. It should not be overlong and must be relevant to your practical. So, for example, if the aim of your practical is to set up an assay for the estimation of human serum albumin levels, you could give a brief introduction which states:

- what human albumin is;
- why you might want to measure it;
- what methods are currently available to measure it;
- which method you are going to use and why.

An introduction that gives a lot of theoretical background but fails to relate it to what you are going to do is incomplete, and will not get good marks. Bear in mind that source material should be acknowledged in the text, for example, 'Bloggs *et al.* (1995) developed a method for ...'.

Materials and Methods

It is not necessary to give a detailed list of materials used unless you have specifically been asked to do so. Instead, you can usually incorporate this into the text. Again, be guided by your tutor on this aspect. In this respect, a report of a single practical differs from a project report, where you would be expected to give details of all the chemicals used and their suppliers (see Chapter 5).

Most often, you will have been given a method sheet to follow by your tutor. Many tutors will allow you simply to insert this sheet into the practical report, rather than slavishly copy it out again. However, instructions are usually written as orders, for example:

> Prepare smears of whole blood as shown by a demonstrator.
> Allow these smears to air dry and then treat as follows:
>
> (a)   Fix in methanol   6 minutes;
>
> (b)   Stain with Jenner   6 minutes;
>
> (c)   Stain with Giemsa   9 minutes;
>
> (d)   Rinse in tap water.

By convention, the Methods section is always written in the past tense: that is, you say what you did, or what was done. This does not mean that you have to rewrite the instructions; you simply have to say: 'The experiment was carried out according to the schedule (attached)'. You must, of course, record any changes that were made during the course of the experiment, either by your tutor or (possibly accidentally) by yourself. So, if you got confused and stained the slide in Jenner for nine minutes instead of six, then you need to say: 'The experiment was carried out according to the schedule (attached) except that slides were stained in Jenner for nine minutes, by mistake'. It is very important to record any changes, however trivial they seem, as this may help you explain any unexpected results. Remember also, that others should be able to repeat your experiment.

You should always record the concentrations of the reagents that you have used, if this has not been recorded in the instructions. Most departments insist that you use the Système Internationale (SI) units. This is a standardized European system of units.

The 'base units' you will come across most frequently are shown in Table 4.1.

Units of mass

The SI unit of mass is the kilogram (kg). In biology, you will usually be dealing with much smaller amounts than a kilogram, and the unit that is most often used here is the gram. You may even use amounts that are smaller than the gram. It is essential that you understand the relationships between these amounts. Have a look at Table 4.2, which explains the symbols used and what they mean. Make sure you know the relationship of these figures to the gram and to each other.

Table 4.1 SI units*

Physical quantity measured	Base unit	SI abbreviation
Amount of substance	mole	mol
Length	metre	m
Mass	kilogram	kg
Time	second	s
Thermodynamic temperature	kelvin	K
Electric current	ampere	A

* Note that the derivatives of these are commonly used in biosciences. These are most commonly smaller than the base unit.

Table 4.2 The kilogram and associated units

Unit	Symbol	Definition
kilogram	kg	$1\,kg = 1000\,g$
gram	g	$1\,g = 10^{-3}\,kg$
milligram	mg	$1\,mg = 10^{-3}\,g$ or $1\,mg = 10^{-6}\,kg$
microgram	μg	$1\,\mu g = 10^{-3}\,mg$ or $1\,\mu g = 10^{-6}\,g$ $1\,\mu g = 10^{-9}\,kg$

Units of volume

The SI unit of volume is the m^3. However, this volume is far too large for biologists to use meaningfully, so we tend to use the following: mm^3, cm^3 and dm^3. Some of you will be more familiar with using other measurements of volume such as the litre (L), the millilitre (mL) and the microlitre (μL). Table 4.3 shows the relationships between these two sets of units:

The relationships between dm^3, cm^3, and mm^3 are shown in Table 4.4.

Table 4.3 SI and equivalent units of volume

SI unit of volume	Equivalent unit
decimetre^3 (dm^3)	litre (L)
centimetre^3 (cm^3)	millilitre (mL)
millimetre^3 (mm^3)	microlitre (μL)

Table 4.4 The relationships between dm^3, cm^3 and mm^3

SI unit	Definition
dm^3	$1\,dm^3 = 10^3\,cm^3$
	$1\,dm^3 = 10^6\,mm^3$
cm^3	$1\,cm^3 = 10^3\,mm^3$
mm^3	$1\,mm^3 = 10^{-6}\,dm^3$
	$1\,mm^3 = 10^{-3}\,cm^3$

Centrifugation

When describing centrifugation, you should always report the relative centrifugal force (RCF) applied to the sample, rather than the speed at which the rotor spins (in revolutions per minute or RPM). This is because the RCF, which is the measurement of the acceleration, applied to a sample within a centrifuge and measured in units of gravity (times gravity or $\times$ g), depends on the radius of the rotor in the centrifuge as well as the speed at which the rotor spins. The relationship can be calculated as RCF $= 1.118 \times 10^{-5} r N^2$ where r is the radius of the rotor in cm and N is the rotating speed in revolutions per minute. Therefore, if you want someone to be able to repeat your experiment, you must either give the RCF, or, if you give the RPM, you must also give the radius of the rotor; for example: 'the cells were centrifuged at 400 $\times$ g for 20 minutes'.

Results

Throughout the practical you should be recording all observations and measurements in your notebook. In your report you should summarize these results in a form that can be easily understood. Pay particular attention to any instructions you have been given by your tutor, who may specifically ask that a set of results be recorded graphically or in tabular form. Most commonly, quantitative results are summarized as a table or a graph. In addition, there are several forms of graphical summary and, unless you have a particular instruction, you should choose the most appropriate form. You should never record the same set of results in more than one form: there is absolutely no need to do this, and it just involves extra work.

Here are some guidelines to help you choose which form to use:
Use a **table** if:

- you have several types of data to display;
- the number of data points is limited, for example, to six or fewer;
- you wish to display numbers to several decimal points.

Use a **graph** if:

- you have only one or two variables but a lot of data points;
- you want to show the relationship between two variables.

A **histogram** is useful if you have a single variable. For example, if you have counted the number of apples that have fallen off a tree in six time periods in September (see Figure 4.2).

If you have counted the apples that have fallen on each day in September, then a line graph might be more appropriate (Figure 4.3).

Other graphical forms include pie charts, scatter plots and area charts. If you are producing your graph using a program such as Microsoft Excel, it is relatively easy to try out the same data in several formats, and see which one suits your data best (assuming your graphical output has not already been specified by your tutor). You will also be able to perform some simple statistical manipulations using Microsoft Excel.

Whichever way you choose to summarize your data, remember that tables and figures (which include graphs, charts, histograms, photographs, drawings

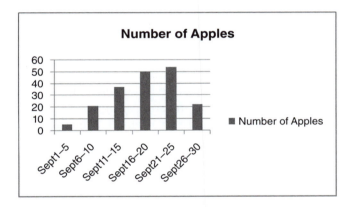

Figure 4.2   Histogram showing the number of fallen apples during September

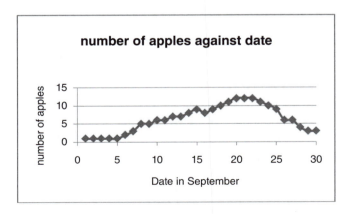

Figure 4.3   Line graph showing the number of apples fallen from a tree in September

and diagrams) should be numbered separately and sequentially. All tables and figures should be referred to in the text. So, for example:

> The results of determining the numbers of apples fallen in September are shown in Figure 1. It can be seen that. ...

The figure or table should have a title which clearly says in bold what it is. For example: **Figure 1: A graph of glucose concentration (in mmol dm^{-3}) against temperature.** Axes on graphs should also be labelled concisely and accurately, with units where applicable.

Calculations

Where you have been required to perform a quantitative procedure, for example, to measure the concentration of an analyte in a sample, you should show how this has been calculated. However, if the same calculation has been applied to a set of figures, you do not need to show each calculation: just provide a specimen calculation so that your tutor can see that you have done this correctly (or not as the case may be!). Bear in mind that your data may also be amenable to statistical analysis. This may be limited to calculations of standard deviations, or may be more sophisticated, using statistical analysis to compare data sets.

Remember always to include units, for example, when recording variables such as concentrations, time, temperature, electric current etc.

Discussion

This section is where you discuss your results. Do not make the mistake of thinking that this is just a restatement of your results. This is where you try to explain what the results are about and to compare them with what was expected, based on known facts from the literature. If your results were unexpected, then you need to reflect on what might have caused these unexpected results. There is a tendency for some students to list anything that might possibly have introduced errors, without any evidence that this has actually contributed to errors. For example, a student might write 'errors could have been introduced by pipetting', but you must consider if there is any evidence that this was the case. On the other hand, if you were conducting an experiment involving enzymes, and you found that the temperature of the water bath was ten degrees cooler than it should have been, then this might explain some odd results.

Your discussion should also attempt to explain what you might do to improve the experiment, or, indeed, what you might do further to investigate the subject. If your results confirm published work, then say so; if they refute published facts, then try to explain why – your conditions or your reagents might have been slightly different.

Conclusion

In this section you will first need to look back at the Aims, and then state whether or not these were fulfilled, and in what way. For example, if the aim was to determine the concentration of glucose in a sample, then the conclusion will be that the concentration of glucose in the sample was x mg $100\,cm^{-3}$ or x mmol dm^{-3}, depending on what units you were using. If the aim was to test the effect of grazing on grass species, then state what the effect was (for example, grazing reduced/increased the number of species in the test area).

Last word on the full laboratory report

Remember that when your written report has been marked, you should read any comments carefully and act on the feedback received so that you do not make the same mistake again. Also, you could also reflect on the skills that you have acquired by undertaking the practical work and by having written the report.

Other forms of report

So far we have discussed how to write up a formal, full, practical report, with a view to getting it assessed. This is still the traditional way to communicate what you did, why you did it and what you found out. If you are completing a much longer practical work, for example, an MSc or a PhD thesis, you may not feel the need to write up each experiment in the full manner. However, it is essential in this case that you keep a practical book, sometimes called a log book. In this you must record, on a daily basis, what you did. You will need to state the reason for doing it, give a very detailed method (including details of reagents and their suppliers), the results, some reflection on what the results mean and what you need to do next. There is nothing worse when writing up a thesis, or even a paper for a journal, to find that some essential details are missing, or that you really cannot remember the reason for doing what you did. Always use a book or file for doing this as scraps of paper get very easily lost. There is more about project work in Chapter 5.

Another form of report occurs when you are conducting some laboratory work, perhaps testing clinical samples, and you need to report the results in brief. You must learn to summarize: a doctor, for example, will not need full details of your methodology when you report results from a patient. The results will inform the diagnosis and subsequent treatment. The doctor will need to know what tests were performed and details of parameters measured that were outside the normal range. Depending on your level of expertise and seniority, the doctor may also want some suggestions for a differential diagnosis. Again, in this case, you will need to be able to summarize to give the major points. The art of summary is a valuable skill and you may be required to use it in other assignments. For example, they may provide you with a scientific journal paper and get you to produce your own abstract of the work. You could do this by yourself: find a journal paper, read it, produce the abstract and then compare yours with the 'official' abstract. You will need this skill when writing an abstract for your project (Chapter 5).

Another way you can practise this, though not related to practical work, is to summarize a novel that you have read in, say, 300 words. Get a friend to read the summary and see if he/she can recognize the novel.

Further reading

Jones, A., Reed, R. and Weyers, J. (2007) *Practical Skills in Biology*, 4th edition. San Francisco: Benjamin Cummings.
Reed, R., Weyers, J., Jones, A. and Holmes, D. (2007) *Practical Skills in Biomolecular Science*, 3rd edition. San Francisco: Benjamin Cummings.

5
The Project Report

About this chapter

In this chapter we will discuss how the project report relates to laboratory reports, and describe in detail the individual sections which make up a project report.

Most taught programmes or courses in the biosciences involve the completion of a project. This may take the form of a written report of a series of laboratory experiments, or a review of literature leading to a written dissertation. The skills and processes you require for this exercise will be described in this chapter.

Introduction

The list in Table 5.1 identifies the process of producing a project report in a stepwise manner. On your journey through this process you will be gaining a variety of skills which are of use in the world of work. They are often referred to as 'transferable skills'. Thus, some of the descriptions elsewhere in this book will be of help to you, for example, searching for information, précis of information, sourcing, citing and organization of references, planning and carrying out laboratory experiments, recording and analysing data, and so on.

A laboratory report normally describes one or a small series of experiments. You have seen in Chapter 4 that the report follows the standard structure of: Aims, Introduction, Methods, Results, Discussion and Conclusion. In addition, you have seen that a scientific article or paper published in a journal has a

Table 5.1   Steps through the project process

- Identify and understand a well-defined current problem in your field.
- Develop appropriate independent practical and analytical skills using proven experimental methods and techniques.
- Recognize the ethical considerations associated with your project area.
- Survey and assess relevant published, peer-reviewed literature.
- Plan and carry out a programme of experimental/research work using proven methods.
- Analyse and present the data in an appropriate way.
- Interpret and discuss the results through comparison and contrasting with other published works.
- Draw conclusions and provide future recommendations.
- Submit a project dissertation or report in an appropriate format as prescribed by your tutor or supervisor.

similar format. Thus the written report for the project follows the same format, including Appendices (plural of 'appendix') and Acknowledgements. The project itself may range from a 'mini-project' at any level during your degree programme, undertaken either individually or as a group, to the 'final-year project', which most students of bioscience will have to complete as part of their degree programme. Conducting and reporting a good project is key to gaining high marks, but, more importantly, is a vehicle for showcasing your 'graduate skills' to potential employers at interviews (see Chapter 8). The projects referred to in this chapter will encompass a written scientific article reporting laboratory findings and a postgraduate project, as the process is similar in each case. Some projects are not based in the laboratory; indeed much of the work in bioscience takes place outside the laboratory. We will also consider how to prepare and report a project which is dissertation based.

How to begin

Planning

The first stage is planning your project. The topic of study may be provided by your tutor. Alternatively, you may wish to identify an area of interest for further study, and begin to develop a 'research question'. Time management is key to successful completion within a set timescale. You should meet initially with your tutor/supervisor to outline the requirements of the project and the scope of the topic. You should then use a diary, calendar or other form of timeline on which to chart your intended work. Try to plan the time required to complete each of your objectives, including time for reading textbooks and journal articles, and time for writing the report, which should take place both during and at the end of the work. The next stage is to formulate and write the proposal.

Proposal

The project proposal should have a working title and outline of the project, including a plan of study, supported by key references. The proposal should be written in the future tense, as you are describing what you plan to do. For example 'I will recruit subjects' or 'I will perform the experiments', or, depending on the guidelines you are given, 'subjects will be recruited' or 'experiments will be performed' rather than the past impersonal 'subjects were recruited' or 'experiments were performed' which is required for writing the final report. Here is a suggested plan for inclusion in the proposal:

- background;
- aims and hypothesis;
- proposed experimental methods;
- the study design;
- statistical analysis;
- references.

Details of **costing** for consumables, consideration of **ethical issues** and an 'ethical review' of the proposed work should be undertaken and included in the proposal. Find out the requirements of your university. Is there an ethics checklist to help you identify issues which may have implications for ethics (see Table 5.2)? Do you need to apply for ethical approval through the university or an independent body? You may need to refer to ethics websites on the Internet, such as the *Integrated Research Application System*, for advice on ethical issues in health and social care, medical or biomedical studies in the UK. The experience of many students is that this part of the procedure takes considerable time, especially if applying externally for permission from an ethics committee to proceed. A **risk assessment** should be included in the proposal. There is a requirement that a health and safety risk assessment and COSHH analysis should be undertaken before beginning your experimental work (Figure 5.1). The acronym COSHH refers to 'Control of Substances Hazardous to Health'. You should identify all substances used in your experiments and find out what, if any, hazards are involved with their use. This includes accidental spillage or splashes, swallowing, skin contamination and safe disposal.

Table 5.2 Example of an ethics checklist

		YES	NO	N/A
1	Does your research involve human subjects?			
2	Does your research involve using NHS patients, staff or resources?			
3	Is the size of sample proposed for any group enquiry larger than justifiably necessary? Have you done a Power Analysis?			
4	Will any lines of enquiry cause undue distress or be impertinent?			
5	Has any relationship between the researcher(s) and the participant(s), other than that required by the academic activity, been considered?			

Table 5.2 (*Continued*)

		YES	NO	N/A
6	Have the participants been made fully aware of the true nature and purpose of the study? If NO, is there satisfactory justification (such as the likelihood of the end results being affected) for withholding such information? (Details to be provided to the person approving the proposal.)			
7	Have the participants given their explicit consent? If NO, is there satisfactory justification for not obtaining consent? (Details to be provided to the person approving the proposal.)			
8	Have the participants been informed at the outset that they can withdraw themselves and their data from the academic activity at any time?			
9	Are due processes in place to ensure that the rights of those participants who may be unable to assess the implications of the proposed work are safeguarded?			
10	Have any risks to the researcher(s), the participant(s) or the university been assessed? If YES to any of the above, is the risk outweighed by the value of the academic activity?			
11	If any academic activity raises questions of legality, is there a persuasive rationale which demonstrates to the satisfaction of the university that: (i) the risk to the university in terms of external (and internal) perceptions of the worthiness of the work has been assessed and is deemed acceptable; (ii) arrangements are in place which safeguard the interests of the researcher(s) being supervised in pursuit of the academic activity objectives; (iii) special arrangements have been made for the security of related documentation and artefacts; (iv) storage of research data is secure?			
12	Have the ethical principles and guidelines of any external bodies associated with the academic activity been considered?			

Writing the report

Although project formats vary from one university to another, in most cases you will find that your project has to contain the following major sections, following the same format as that recommended for submission of original articles to scientific journals. Nevertheless do check the requirements for your own university or the 'instructions to authors' if you are considering publication of your work in a scientific journal.

Hazard	Action
Eye contact	
Inhalation	
Skin contact	
Ingestion	
SPILLAGE	
Non-acidic	
Acidic	
Water immiscible	
Solid	
DISPOSAL	
Liquid	
Solids	
Glass and Sharps	
	HAZARDOUS MATERIALS MUST NOT BE REMOVED FROM THE LABORATORIES

Figure 5.1   Extract from a COSHH assessment form

The abstract

This is a summary of the work undertaken and the results obtained. It should include a brief background and introduction, aims, a short description of study design and methods, analysis of results and key conclusions. Your writing style for the abstract should be in the present tense, as you are describing what is to come in the full report. It is acceptable to use 'I' or 'we'; for example, 'We report the findings of leukocyte analysis in sickle cell cases'. However, different institutions may have views on this, with some preferring an impersonal form. Always read and follow the guidelines you have been given. It is often recommended that the maximum length is 300 words. Again, find out what your university (or a publisher, if planning to submit the report for publication) requires. Some abstracts include sub-headings, whilst others are composed of a series of short paragraphs. Note that the abstract should not include references, tables or figures. Try not to use abbreviations, or keep them to a minimum. The abstract is an important section, as it can be reproduced alone, for example, in handbooks of scientific conference proceedings. This should be the last section you write, having completed the remainder of the report.

An example of an abstract is shown in Figure 5.2. This abstract is taken from a short paper describing a study of sickle cell disease patients. Note the conciseness and detail of the information provided.

Homozygous sickle cell disease (SCD) is characterised by increased soluble P-selectin (sP-selectin, suggesting increased platelet activation), and high non-transferrin bound iron (NTBI, reflecting iron overload, possibly due to blood transfusion). Hypothesising a relationship between these processes, we measured both markers in 40 SCD patients and 40 age/sex/race matched controls, finding (as expected) increased levels of each marker in the patients (both $p<0.001$), but more pertinently a significant NTBI/sP-selectin correlation ($r=0.52$, $p<0.001$). Both indices were increased in the blood of 15 recently-transfused patients compared with 25 three-month transfusion-free patients ($p<0.001$), but only sP-selectin was higher in present sickle crisis ($p<0.001$). We suggest that increased NTBI associated with blood transfusion iron overload in SCD may promote platelet activation.

From Blann, A. *et al.* (2007) Increased levels of soluble P-selectin correlate with iron overload in sickle cell disease. *Br.J. Biomedical Science*, **64**(3), 124–127. Extract reproduced with permission of *British Journal of Biomedical Science*.

Figure 5.2   Example of an abstract

Introduction

An investigation of the background to your study and up-to-date literature survey should be carried out. You should write this section in the form of a critical review of the topic in such a way as to place the current project work into context. The introduction is organized as a series of paragraphs or subsections. The writing style is concise and densely referenced. It should begin with a broader view of the topic, however, the focus should narrow in later sections as specific questions arise and develop into the aims of the research topic.

Aims and objectives

You must state a clear aim or set of aims. This helps the reader to focus on what you are attempting to achieve in performing the project. One way to phrase the aims and objectives is to ask yourself two questions; 'what do I want to find out?' and 'how will I go about achieving this?' You may wish to state your hypothesis here.

An example of the aims, objectives and hypothesis of a study investigating a link between sickle cell disease and leukocytes is shown in Figure 5.3.

Aim: To investigate whether there is a link between sickle cell disease and leukocyte markers.

Objectives

1.   To establish the presence of a range of leukocyte markers in normal healthy controls and in patients.

2.   To confirm the presence of sickle cells in patients using cellulose acetate electrophoresis.

3.   To evaluate the immunophenotype of leukocyte markers in patients' blood.

Null hypothesis: There is no difference in leukocyte markers between normal controls and patients of sickle cells disease.

Alternative hypothesis: There is a difference in leukocyte markers between normal controls and patients of sickle cell disease.

Extract from unpublished PhD research project 2009, Manchester Metropolitan University.

Figure 5.3   Example of the aims, objectives and hypothesis of a study

Materials and Methods

This section contains full details of the methods you have used in attempting to achieve the aims of your study. You should always use the past tense in the passive mood (see Chapter 1), for example 'the samples were centrifuged' rather than 'I centrifuged the samples'. Details of dilutions, concentrations of solutions, units of measurement and other experimental procedures are identical to those described for laboratory reports in Chapter 4. It is common for some experimental procedures to use pre-prepared kits, gels or other consumables, which come with instructions for use from the manufacturer. Do not fall into the trap of copying directly from a manufacturer's recommendations or a laboratory procedure schedule. There are two main reasons for this: first, because you are describing what you, yourself, actually did, rather than what others stated should be done; second, because you will be tempted to use the present imperative tense, as it is described by the manufacturer. The detail should be sufficient to enable the reader to repeat the experimental work without requiring further information. An alternative way would be to cite a reference for the methods you have used, providing you followed that method exactly or stated any changes to the published method. Whilst this is often found in published articles, you may need to check with your supervisor whether this approach is appropriate for your own methods section. Ethics should be referred to here. State whether ethical permission was applied for and granted, or was considered and found unnecessary. It is not sufficient to ignore ethics on the basis 'there aren't any ethics in my project'.

An example of text and style of a methods section from a short project is shown in Figure 5.4.

This example of method description includes:

- where the study samples were obtained;

- details of ethical approval;

- how the samples were handled and stored;

- details of the method used to analyse the samples;

- details of the statistical analysis;

- details of the numbers of samples required.

Forty consecutive patients with SCD (mean/SD age 35/8 years, 23 males) were recruited from those attending our haemoglobinopathy service. Diagnoses were confirmed by Kontron HPLC. They were classified at the time of venepuncture as being in steady state (currently free of vaso-occlusive complications and a transfusion free interval of <3 months, n=25) or in crisis (acute bone pain requiring opiate analgesia and/or vaso-occlusive events such as acute chest syndrome, n=15). Transfusion history was also recorded, focusing on the previous three months. Forty race, age and sex matched subjects with a normal haematology profile and negative sickle screen test and who had not been transfused for three months were recruited from those attending Out-patient Clinics. Local Ethical Committee approval was obtained, as was informed consent of all participants.

Citrated and serum blood specimens were centrifuged within one hour at 3000 rpm (1000g) for 20 min and the plasma and serum stored at -40°C. Soluble P-selectin was assayed by a standard technique (ELISA, R&D Systems, Abingdon, UK) in citrated plasma, NTBI levels were determined in the serum sample by a bleomycin based assay (18). The principle of the latter, briefly, relies on bleomycin breaking one of the double strands of a sample of DNA to which free iron in the serum sample binds. This in turn catalyses the production of malonyldialdehyde by hydrogen peroxide and ascorbate, which can be detected in complex with thiobarbituric acid at a wavelength of 405 nm.

The relationship between soluble P-selectin and NTBI concentration, and transfusion status was assessed using the Mann-Whitney U test and Spearman's Rank correlation method. With a non-normal distribution (Anderson-Darling test), data is presented as median and inter-quartile range. The power calculation for our primary hypothesis was based on the expectation of correlation coefficient of a least 0.4, which we take to be meaningful. Accordingly, our power calculation was that 37 data points are needed as these provide the one sided alpha at 0.05 and 1-beta at 0.8 (19). Thus we recruited to slightly in excess for additional confidence. For our secondary hypotheses (that transfusion history and/or clinical status of crisis/no crisis influences the research indices), a sample size of n=30 and n=40 provides the 1-beta = 0.8 and p<0.05 power to detect a 33% and a 30% difference respectively between the research indices. Analyses were performed on Minitab 13 (Minitab Inc, Progress Drive, State College, PA, USA).

From Blann, A. *et al.* (2007) Increased levels of soluble P-selectin correlate with iron overload in sickle cell disease. *Br.J. Biomedical Science*, **64**(3) 124–127. Extract reproduced with permission of *British Journal of Biomedical Science*.

Figure 5.4   Example of text and style of a Methods section from a short project

Results

In this section you should detail the data you have obtained in your work. Where possible the data should be in a summarized form with appropriate statistical analysis. Graphs, figures and tables should conform to specifications provided by the university or guidelines to authors. You must avoid presenting the same data in more than one form, for example, displaying data in a table and using the same data in a histogram. You should provide explanatory text within the results section in order to refer to figures and tables and to describe what each figure or table shows. It helps to think of it as a story you are telling to someone else, but using scientific grammar. The term 'figures' includes pictorial information such as graphs, charts, histograms, photographs, micrographs, drawings and diagrams. All figures and tables must be numbered in the text. Numbering should begin at 1 and include the chapter number. For example, if placing a table in Chapter 3, tables would be labelled Table 3.1, 3.2, similarly Figure 3.1, 3.2 and so on. You can use either the full word 'Figure' or abbreviation 'Fig.', depending on the requirements of your university or a publisher. Whichever you choose, make sure that you stick to the same form throughout your written report. All figures and tables must include a title and a legend. A legend is a short paragraph which briefly summarizes the content, placed directly below the table or figure, separate from the main body of the text.

An example of results showing a written description of a table is shown in Figure 5.5.

Table 1: Levels of sP-selectin, NTBI and number of blood units received in Study Subjects

A. Patients and controls

	Patients (n=40)	Controls (n=40)	*p* value
sP-selectin (ng/ml)	87 (46 - 128)	66 (48 - 104)	<0.001
NTBI (µmol/l)	1.25 (0.25 - 3.2)	Not detected	<0.001

As expected, soluble P-selectin and NTBI were significantly higher in SCD patients compared to the controls (Table 1). With respect to our primary hypothesis, there was a significant correlation between NTBI and soluble P-selectin (r=0.52, p<0.001) (Figure 1). Soluble P-selectin also showed a significant positive correlation (r=0.59, p<0.001) with the number of red cells units transfused.

With respect to our secondary hypotheses, we found a further increase in soluble P-selectin during sickle crisis. In addition, soluble P-selectin was significantly higher in the recently-transfused patients compared to those free of recent transfusion (>3months). This finding paralleled the significantly raised NTBI in transfused patients. All differences are within the limits of our power calculation, indicating minimum risk of types 1 and 2 statistical error.

From Blann, A. *et al.* (2007) Increased levels of soluble P-selectin correlate with iron overload in sickle cell disease. *Br.J. Biomedical Science*, **64**(3) 124–127. Extract reproduced with permission of *British Journal of Biomedical Science*.

Figure 5.5 Example of part of a Results section

Discussion

This section is a measure of your own intellectual input. In the discussion you should analyse the results of your study and discuss them critically in the context of your project and in relation to other, published research. It is a good idea to suggest what future work might be done to extend the study, or the next stage in the research. Ask yourself the question 'what would I have done next if I had more time or resources?'. A common temptation for students is to try to discuss new information, or refer to something outside the project. Have you achieved your stated aims? Have you proved or refuted your hypothesis?

Students sometimes use the discussion to discuss published results but omit to discuss their own results. Another mistake is to regard the discussion as a place to restate the results. The discussion should attempt to explain the results.

References

This is a list of the sources of information you have used and quoted throughout your project report. It will include published journal articles, personal communications, websites, chapters from textbooks and manufacturer's information sheets. Details of how references are quoted in the text and listed in the references section are given in Chapter 2. Please note that, while the use of web-based material as a source of information is often encouraged and can be invaluable, primary sources in the form of journal-based references to the material should be quoted since these have been subject to peer review (see Chapters 1 and 2). It is important that you are able to distinguish this 'primary' information, by which we mean original articles that have been through a review process, from other web-based material that the public can populate. Examples of the latter include Wikipedia, and other uncontrolled public websites.

Dissertation-based projects

What is a dissertation-based project? The basic description of a dissertation might be a 'long essay, which has a contents list and introduction at the beginning and discussion and conclusion at the end'. A project based on a literature review will be reported in the form of a dissertation. This may be as long as a laboratory or field-based project, and must be of equal rigour in terms of the skills you will need to develop.

In planning a **project proposal** for a dissertation-based project you should consider the following:

1. **Project title.** This is provisional to begin with. It should be informative and probably no longer than twenty words.

2. **The 'research question'.** You should write a concise description of the main question to be explored in the project. You should also include two or three subsidiary questions, for example, you wish to study the efficacy of red cell concentrate transfusion in the oxygenation of tissues. The research question is: 'Does red cell transfusion improve the oxygenation of tissues?' The subsidiary question might be: 'How can tissue oxygenation be accurately estimated?'

3. **Abstract.** This comprises a brief outline of the proposed investigation highlighting its main points and the probable findings or outcomes if your research question is valid. You are advised to write this section last. Aim for a maximum 200–250 words.

4. **Aims and objectives of the investigation.** As with an experimental study, the aims and objectives must be clear. A rule of thumb is to list between one and six aims and identify approximately two to ten objectives. The objectives will indicate how you intend to go about achieving your aims. For example 'The aim is to … and this will be achieved by …', that is, 'the objectives are to …'.

5. **Background.** This section should clearly set out the context to your proposed investigation and show that you have sufficient knowledge of the background relating to your work. You should have references to at least two key scientific papers that are directly relevant to your research question, in addition to others that set the scene for the investigation.

6. **Methods.** The method description for a dissertation-based project should be based on the normal procedures for an evidence-based review, describing the approach that you intend to use in your dissertation. You should include definitions; for example, define the subject, intervention outcomes and the approach you will use to search the literature. This will involve thinking about which key words or search terms you will select. The criteria for the inclusion of your selected journal articles in your dissertation should be identified.

7. **Risk assessment.** Identify any procedures that require an assessment of risk. This is performed to ensure the health and safety of the individuals performing the work, for example, working with a computer. Do the procedures already exist? Most common procedures are already assessed for risk, so the information should be available. If

not, you should describe the steps needed to identify any risks before you begin the work.

8. **Ethical issues.** Identify ethical issues and appropriate actions that are required to satisfy the university's requirements (and external requirements where relevant). Note that there may be none, but you need to justify this. An example of an ethics checklist is shown in Table 5.2.

9. **Data analysis and presentation methods.** What kind of data will you collect? Will your analysis be quantitative or qualitative? If appropriate, what statistical and summary (tabular or graphical) procedures will you use?

10. **Timescale.** Produce a time line for the major stages of your project including your literature search, the start and finish dates for your experimental work and for completing the report. It is often easiest to work backwards from the final date. You are strongly advised to begin writing up long before the end of data collection. Do not wait until the end of this activity before you start writing.

11. **References.** List all that are cited and do not list any that are not included, make sure that you adopt an appropriate and recognized format, such as the Harvard form of referencing (see Chapter 2).

Ethical issues to consider when reporting your project work

In addition to completing an ethical review for your experimental work you should be aware of how to report your project in an ethical manner. This includes:

- honesty and integrity in reporting results – do not invent, hide or alter data;

- acknowledgement of others who have contributed to your work, for example, technical staff, sources of funding, collaboration with other universities, a colleague who has read and provided feedback on a draft of your report, sources of materials provided by researchers (for example, cell cultures);

- maintaining confidentiality, for example, personal details of, or results relating to, human subjects. Your data should be stored on specified, password-protected areas of a computer. The data should be disposed of after a specified time, in accordance with the Data Protection Act;

- declaring any conflict of interests, for example, research on a specific pharmaceutical agent which is supplied by the pharmaceutical company. Clearly the company would have an interest in the outcome of results in terms of the future commercial success of the drug, and this could bias the way in which you report the results;

- declaring that the report is your own work. This is normally the case for a student undergraduate project, however, a report for publication may include the names of co-authors who have significantly contributed to the design or data collection/analysis and to the written report. Project reports should include a statement or declaration which is signed by the student;

- be aware of copyright and ownership of intellectual property. Who owns your data? The answer is the body that has funded your project, usually the university in the case of undergraduates, but this may not always be the case if the project has received research funding from elsewhere. Written work must be in your own words. You should revisit Chapters 1 and 2 for guidelines on avoiding plagiarism.

And finally ...

The following is a checklist of common errors in project reports (and shows a very quick way to lose marks for your project!).

1. You have not established a hypothesis.

2. You have duplicated information throughout the report.

3. You have not included a narrative of your results.

4. You have not included legends for your tables and figures.

5. You have provided insufficient discussion of the findings.

6. You have failed to refer to future, or others', work in your discussion.

7. You have not included up-to-date references.

8. You have not included all references cited in the text in the reference list.

9. You have cited your references incorrectly.

10. You have placed too much emphasis on your literature review or introduction.

11. You have described principles of techniques in the method section, rather than the method itself.

12. You have failed to follow guidance on the layout and binding of the report.

Further reading

Control of Substances Hazardous to Health (COSHH) website. Available from: http://www.method-statement-template.com/coshh.html [last accessed 10th September 2009].

Kirkman, J. (1992) *Good Style for Scientific and Technical Writing*. London: E. & F.N. Spon.

Malmfors, B., Garnsworthy, P.C. and Grossman, M. (2000) *Writing and Presenting Scientific Papers*. Nottingham: Nottingham University Press.

National Academy of Sciences, National Academy of Engineering, National Institute of Medicine (2009) *On Being a Scientist: A Guide to Responsible Research*, 3rd edition. Washington: National Academies Press.

Open University (2007) *LDT_4: More Working with Charts, Graphs and Tables* [online]. Available from: http://openlearn.open.ac.uk/course/view.php [last accessed 9th October 2009].

Open University (2007) *T175_7: Presenting Information* [online]. Available from: http://openlearn.open.ac.uk/course/view.php [last accessed 9th October 2009].

6
Scientific Posters

About this chapter

This chapter sets out the purpose of posters in different situations and emphasizes the need for different styles for different audiences. It gives practical advice for producing and presenting posters.

Why use posters?

A poster is a popular and effective method for communicating scientific information and over recent years poster presentations have become an increasingly important part of coursework and, generally, of the academic activity of departments and schools. For many years, poster presentations have been a standard feature of academic conferences, and, of course, they are major tools of commerce and industry to get across information about their products. In fact, posters are a very good and effective way of providing information to others about the work you are doing. However, like all advertising, you have to grab and hold the reader's attention. In a sense, therefore, the style of presentation is at least as important as the information you wish to convey. In this way it differs from, say, an essay or a journal article, where the information you wish to present to some extent takes precedence over the style of the presentation.

The skills you acquire in learning to produce a well thought out and effective poster will be enormously useful in your later career, whether you choose to be a professional bioscientist, or take up another profession. They are also likely to be extremely handy if you become active in your local community, because

Communication Skills for Biosciences   Maureen M. Dawson, Brian A. Dawson and Joyce A. Overfield
Copyright © 2010, John Wiley & Sons Ltd.

they will allow you to get your message across quickly and effectively to members of the general public.

Posters are ideal for communication in a variety of situations. This is because the poster format allows for concise and succinct portrayal of key information, in a relatively 'small' physical space. A poster can be transported to different venues for display. Less frequently, a poster can be electronically displayed using a computer and data projection.

There are several things to consider if you are going to produce an effective and attractive poster. It is clearly sensible to spend time before you begin to think about what you want to say and how you want to say it, particularly as the layout and graphics are important.

When to use a poster format

Some examples of the use of posters include:

- at a national or international scientific conference;
- at a local science meeting, for example a subject discussion group;
- as an assignment in your programme of study, individually or in a group;
- to communicate to the public, for example in a school or doctor's surgery waiting area.

In the world of science, recent research findings are communicated, in particular by one of three methods: a written and peer-reviewed journal article; a peer-reviewed oral presentation, usually called a 'paper', at a conference; and a peer-reviewed poster also presented at a conference. The term 'peer review' means that other researchers in a similar field of work have read and commented on the information being communicated (see Chapter 2), for example the results of a study or description of an interesting clinical case in the medical arena. Thus, peer-reviewed posters are displayed at conferences attended by hundreds, and possibly thousands, of delegates. They may also subscribe to a journal relevant to their subject area and can read the 'conference proceedings', that is, a compilation of abstracts in which the content of the poster is briefly described. Posters are a well-regarded means of scientific communication. They are often the first method used to communicate information to a wider scientific audience. In fact, most scientific meetings include time for the display of posters and you, as the presenting author, would be expected to be present during this session. The reason for this is that it gives you the opportunity to discuss ideas and results with your peers. Therefore it is advisable that you should know your audience. Most people walking past your poster will, at best, skim-read the title. You need to catch the attention of the most important people.

Designing a poster

One of the first things you need to find out is how much space do you have in which to display it. If you have no information about space it is usually safe to assume approximately a one metre squared space. The next stage is to identify what information you can include and what you need to leave out. It is not always necessary to include the complete findings of a large project; you may display a proportion of your work, in which case this would be stated on the poster, usually near the bottom edge. To determine the specific size require-ments ask your university tutor, supervisor on, if for a scientific conference, consult the conference organizers or their website for information. The most common sizes are A1 and A0 (see Table 6.1). As you can see from the table, the size of each paper in the series is half the size of the one above. The space available determines how the content will be organized. There are two choices of orientation – portrait (long and slim) or landscape (wide).

Table 6.1 Standard 'A' series paper sizes

Size	Dimensions (mm × mm)
A0	841 × 1189
A1	594 × 841
A2	420 × 594
A3	297 × 420
A4	210 × 297

Remember that your poster needs to grab the attention of 'browsers', but needs also to have valid scientific content. In designing an effective poster, you should take account of the following points:

1. The poster should be succinct and well organized.

2. The poster should have a clear 'take-home' message?

3. The poster should be attractive and aesthetically pleasing.

4. The poster should be easily readable from a distance of at least one to two metres.

5. The poster should be focused on the science.

6. Use short sentences and paragraphs of no more than twenty lines.

7. Use lists whenever possible.

8. Mix text and graphics.

9. Choose the colours carefully.

10. Choose the type font, size and spacing carefully.

Use of colour

Colour gives your poster impact and adds to the interest value. Try to use two to four colours, but no more. Colours should match well together; avoid clashes. Primary colours – blue, red and green – are good choices. The text should be in strong contrast with the background colour, thus a light background, for example light blue, requires a dark colour of text, such as dark blue or black. Try experimenting with different backgrounds and texts on your computer to get an idea of what looks effective, though be aware it may look different when printed. Placing an image as part of the background does not tend to be effective as the result is too fussy, and detracts from the text.

What should my poster include?

There are two important elements to a poster: style and content. The style of a poster for the communication of experimental findings is fairly well prescribed. It should include sections that are similar to a laboratory practical report (Chapter 4) or a project report (Chapter 5). Table 6.2 lists the sections suggested for a poster the purpose of which is to communicate results of experimental studies, and which is aimed at the scientific community. It is also sometimes

Table 6.2 Posters checklist

Section	Tick when done
Title	
Names of authors or contributors	
Address of authors or contributors	
Keywords	
Brief introduction	
Methods used	
Results	
Discussion/Conclusions	
References	
Acknowledgements	

useful to provide A4 handouts of the poster or abstract and include your name and email/address so interested readers can contact you.

The balance of text, images and space is recommended to be approximately 2:4:4. You are therefore aiming for roughly twice as much space as text.

The title. Think carefully about the title. It should accurately reflect the information yet be reasonably concise. Choose words that will be relevant for use in a keyword literature search. Avoid phrases such as 'an investigation of …' as this is unnecessary information. The font size of the letters in the title should be easily readable 'across a room' or more specifically at a distance of approximately three metres. Select a minimum font size of 72 to produce a title approximately five centimetres high (see Table 6.3 for example font sizes). Imagine you are in a venue faced with a number of posters. You need to be able to select the ones you want to read without having to peer closely at each one to find out the title! Choose a sans serif font such as Verdana, Arial or Tahoma as these will be clear from a distance.

Table 6.3 Suggested font sizes

Content	Reading distance (m)	Size (point)	Style of type
Title	3–4	72–96	**Bold**
Authors	3–4	54–72	**Bold**
Main headings	2+	48–54	**Bold**
Sequence numbers	2+	36	Regular
Main text	1+	24–28	Regular, single spacing
Detail text	1+	18–24	Regular

Examples of titles. Titles should be succinct and clear. If submitting your poster (and the accompanying abstract) to a scientific conference you should check with 'instructions' for a maximum word count for the title. For example 'Chromatographic measurement of alcohols' is better than 'An investigation of gas chromatographic systems for the measurement of ethanol, methanol and ethylene glycol' and 'The utility of an automated blood cell counting system in clinical paediatric haematology' becomes 'Paediatric blood cell estimation'.

Names and addresses of authors or contributors. Include yourself and others who have contributed significantly. The first name listed is the 'first author'. If you are submitting to a scientific conference, this is usually the person who is the 'principal investigator' of the research and whose responsibility it is to submit the abstract to the conference organizers, and attend the conference to present the poster. The names should normally follow this format: Johnson, B.G., Smith, C.L. and Roberts, J., although a less formal format such as 'Maureen M. Dawson and Joyce, A. Overfield' can also be used. And the

address of each author is placed below the names, each address being indicated by a number superscript unless all authors work in the same department (see Figure 6.1). The addresses are provided so that it is clear where the research has taken place, and so that the authors can be contacted for future enquiries. Logos showing affiliations are useful. Placing the logo on the top corners or along the bottom of the poster will show which institution(s) or professional bodies were involved in the project or study. You could include your university or college logo. A passport-size photograph of yourself is often required – this helps to identify you as the 'presenting author' in a busy poster session at a scientific conference or university.

Keywords. Selecting the right keywords is important as they are used in searching literature databases. Choose three to four relevant words to facilitate the process of information retrieval and selection by others who are researching or studying in a similar topic. Keywords for the above titles could be 'chromatography' and 'alcohol', and 'paediatric' and 'blood cells' respectively.

Introduction and background. This section should describe the reasons for undertaking the study/research, and should be supported by a few key references. (Using the Vancouver/numeric system for references may be acceptable as space

Development and implementation of a policy for delivering effective feedback to students

MAUREEN M. DAWSON[1], JOYCE A. OVERFIELD[2], CAROL AINLEY[2], ALAN FIELDING[2], ROD CULLEN[1] AND MICHAEL COLE[2]

[1] Learning and Teaching Unit, Manchester Metropolitan University (MMU), St. Augustine's Building, Manchester M15 6BY
[2] School of Biology, Chemistry and Health Science, MMU, Oxford Road, Manchester M1 5GD

INTRODUCTION

Provision of timely, high-quality feedback on assessments has been identified as supporting student learning (Gibbs and Simpson, 2004, Juwah et al., 2004). The two previous National Student Satisfaction Surveys of level 3 students indicated feedback as an area which could be improved in clarity and detail of comments and in timeliness. Local surveys of level 2 students at MMU have produced similar findings.

Currently, all modules within the Biology and Health Sciences divisions at MMU are delivered within a VLE (WebCT Vista). Use of the VLE, combined with feedback proformas could therefore provide a means for delivering rapid feedback to students, particularly if staff confine their feedback to those learning outcomes being assessed at the time.

AIMS

The aims of the project are to develop, evaluate and embed a School procedure for delivering effective, relevant and high-quality feedback on assignments.
The objectives are to:
• survey the extent of use and primary purpose of feedback proformas used within the School;
• devise, trial and evaluate the use of feedback proformas which are explicitly linked to the learning outcomes of different types of assignments.
• transmit feedback proformas to the students through the VLE.
• embed the use of feedback proformas, if deemed successful, within all modules in the School.

METHODS

1. Academic staff were asked to provide a copy of any proformas or marking schedules used in assessing student assignments.
2. Academic staff (eleven in total) took part in a structured interview in order to clarify current practice regarding the use of feedback proformas. The questions used are shown in Table 1.
During the interview staff were also asked to consider what would be their expectations of the learning outcomes for different types of assignment at different levels of the programme.

Table 1: Focus group with academic staff: questions
• Do you use a proforma when marking student assignments?
• What is the primary purpose of the proforma that you use?
　• To make marking easier?
　• To make marking more consistent?
　• To provide evidence for audit purposes?
　• To provide useful feedback to students on their assignments?
• Do you provide generic feedback on assignments? If yes, do you use templates?
• Do you provide an assignment brief? If yes does this brief make clear to students the learning outcomes of the assignment, the mark allocation, the skills being tested and the link to grade descriptors?

RESULTS AND DISCUSSION

More than thirty individual proformas are currently in use within the Biology and Health Sciences divisions at MMU. However, there is considerable overlap between them. There is therefore scope for using proforma templates, which can be adapted to suit individual tutor need.
The structured interviews of academic staff showed that, often, assessment proformas are used to:
　• make marking easier;
　• make marking more consistent; and
　• provide evidence for audit.
However, most staff agreed that the primary use of proformas should be to provide useful feedback for students.
The survey of the proformas frequently demonstrated that the language of the proformas was not 'student friendly' and may, indeed, be confusing.
This exercise has raised awareness amongst staff of the need to simplify proformas so that students can readily access their message, rather than simply register a mark.
Staff have identified generic learning outcomes for different assignments at different levels. One example is shown in Table 1.

ACKNOWLEDGEMENTS
This work is supported by a grant from the Higher Education Academy subject centre for Bioscience.

Figure 6.1 Poster displaying the results of a project. This poster was presented at the Science Learning and Teaching Conference (2007) (see http://www.bioscience.heacademy.ac.uk/hosted/sltc/2007.asp) and can be downloaded from http://www.bioscience. heacademy.ac.uk/ftp/events/sltc07/papers/p10dawson.pdf

is limited but do check requirements for references with your university tutor or conference organizers – see Chapter 2 for details of referencing and advice.)

Methods and results. The methods should be described concisely, perhaps using bullet points. A key reference may be used to refer to the method, if it is already published elsewhere.

Results should be presented as tables and figures. The term 'figures' includes pictorial information such as graphs, charts, histograms, photographs, micrographs, drawings, diagrams. Each must have a clear heading and be numbered consecutively. Data or other measured values should be reported in the same way as a full written report or article and include SI units (see Chapter 4). Summary statistics and analysis should display error bars, standard deviation, standard error/mean/median values. In short, the rules for methods and results in a project report also apply to the presentation of your poster.

Discussion and conclusions. This section should summarize the key findings. You may choose a few key phrases or bullet points. Alternatively a short paragraph of text may be effective. Whichever you choose, you should think about what is the 'take home' message you wish to get across.

Acknowledgements and references. Acknowledgements should be included in the same way as in a full project report or thesis (see Chapter 5). A selection of key, recent references should be placed at the bottom of the poster. You could consider using a smaller font size to save space. Affiliations, that is work supported by or in collaboration with other universities or research groups, can be shown by displaying the logo of the other party.

Printing your poster

There are several options depending on how much money you have available to spend. You can choose to print your poster on A4 or A3 sheets of paper, preferably glossy good quality paper, then mount it on cardboard. This means you will have to pin up a number of smaller sections to make up the full display. The benefits to this option are that it is relatively cheap, and the poster sections are easily transportable as they fit into a briefcase or suitcase. Alternatively you can contract the printing to a commercial printer or use your university facilities. This comes with a small cost, depending on size and quality. Expect to pay more if you want the poster to be laminated, which makes damage during transport less likely. You can select to print the full poster in one sheet, sized, for example A0 or A1. Posters A0 sized are printed on paper which is 910 mm wide and 1220 mm long. Most students use this format, especially since the creation of posters using a single Microsoft PowerPoint slide (set to the poster size) began to be popular. The large poster is then transported to the venue for display rolled up in a plastic or cardboard cylinder.

Examples of posters

Figure 6.2 shows an example of a poster with many good features, including:

- the layout is such that you can see the order in which to read;
- the details of addresses and a logo are provided;
- each box has a bar (coloured in the original) to show clearly where the sub-headings are, making it easy to see 'where you are' in the information;
- aims are described briefly;
- results and units of measurement are displayed;
- statistical analysis is provided;
- there is a good balance of images and text;
- all relevant aspects of a good poster are present, though using a sans serif font and perhaps a larger font size for the title would make it even better.

Figure 6.3 shows an example of a poster completed by a group of students working together on a case study for a specific disease. In this type of poster you are collecting and displaying information which already exists, as opposed to communicating findings of experiments or research. The poster is used to communicate information in a concise and pictorial manner.

The good points:

- provides a clear layout;
- uses concise phrases;
- provides plenty of information;
- consistent with font sizes for headings.

Areas for improvement:

- lacks colour (lacking in the original) and images;
- lacks use of tables or figures;
- references take up too much space;
- lacks space, is too crowded.

Development of a staphylococcal biofilm model for testing antimicrobial agents

Student Name, MSc in Medical Microbiology

School of Biology, Chemistry and Health Science

Manchester Metropolitan University

Chester Street, Manchester M15 5GD

Manchester
Metropolitan
University

Introduction

A biofilm is defined as a collection of microbial cells which is difficult to remove by gentle washing from a surface (Donlan, 2001). *Staphylococcus aureus* and *Staphylococcus epidermidis* are associated with many biomaterial-related infections. Typically, these infections are associated with biofilm formation. Bacteria in biofilm are thousand-fold more resistant to treatment with antibiotics than the same bacteria when grown in planktonic state (Gilbert *et al.*, 1997). As a result, biofilm-related infections are inherently difficult to treat and to fully eradicate with normal treatment regimens. Here, conventional 96-well microtitre plates coupled to Crystal Violet assay, XTT assay and TTC assay were used for the development of staphylococcal biofilm model to test essential oils and biocides.

Aim

The aim of this project was to determine the most appropriate susceptibility testing techniques for *S. aureus* and *S. epidermidis*.

Methods

S. aureus (NCTC 6571) and *S. epidermidis* (NCTC 11047) were used as test bacterial strains. The standard curve for viable count (cfu ml-1) and Crystal Violet (CV) (optical density) was drawn where five different MacFarland (MF) standard concentrations (4, 2, 1, 0.5, and 0.25) were used for *S. aureus* and *S. epidermidis*. Three different published assays were used to quantify the biofilm formation abilities of *S. aureus* and *S. epidermidis* in 96-well microtitre plates (i) the XTT [2,3-bis (2-methoxy-4-nitro-5 sulfophenyl)-2H-tetrazolium-5carboxanilide] reduction assay (Thein, *et al.*, 2007), (ii) the TTC (2,3,5-triphenyltetrazolium chloride) reduction assay (Juda *et al.*, 2008), and (iii) absorbance following staining by CV dye (Li *et al.*, 2003). The optical densities (OD) were determined. To test the ability of these assays to quantify antibiofilm activity, the biofilms were exposed to two biocides; hydrogen peroxide (H_2O_2) and hypochlorite (HCLO) at concentrations of 3% to 0.005859% and two essential oils; clove bud and lemon grass at concentrations of 2% to 0.003906%. The visible Minimum Inhibitory Concentrations (MICS) were determined.

Statistical analysis

Statistical analysis (one-way Analysis of Variance, Tukey test) was performed with the Minitab Version 15 software package, with P values < 0.001 considered to be statistically significant.

Results

Fig. 1: i Light; ii medium; iii heavy growth of biofilm.
a *S. Aureus*; **b** *S. epidermidis*
in 96-well microtitre plate after CV staining.

Results (cont.)

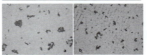

Fig. 2: Microscopic observation of biofilm of **a** *S. aureus*, **b** *S. epidermidis* in 96-well microtitre plate after CV staining (× 40).

Quantification of biofilm by CV assay and viable count at different MF are shown in Figures 3 and 4.

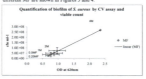

Fig. 3: Mean log of cfu of viable counts and mean of O.D. of CV.

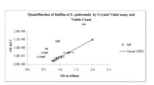

Fig. 4: Mean log of cfu of viable counts and mean of OD of CV.

Quantification of biofilm by XTT assay, TTC assay and CV assay are shown in Figures 5 and 6.

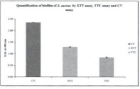

Fig. 5: Mean of OD of CV, XTT and TTC.

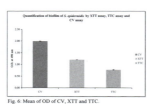

Fig. 6: Mean of OD of CV, XTT and TTC.

Results (cont...)

Determination of visible MICs for clove bud, lemon grass, H_2O_2 and HCLO are in Table:1

Organism	H_2O_2	HCLO	Clove bud	Lemon grass
S. aureus	3%	3%	2%	2%
S. epidermidis	3%	3%	2%	2%

Table 1: MICs for H_2O_2, HCLO, clove bud and lemon grass are presented in (%).

Discussions

The four MF standard concentrations showed the highest viable count, and highest OD of CV staining for both *S. aureus* and *S. epidermidis*. The data were reproducible for this starting concentration as it showed low error. Results for the CV assay were more reproducible and had the strongest correlation to viable counts for both *S. aureus* and *S. epidermidis*. The CV assay showed higher OD than the XTT assay and the TTC assay (p<0.001). The CV assay can be used to quantify biofilm inhibition as this gives no information on viability since both live and dead cells will take up the stain. Any antimicrobial activity on preformed biofilms must be measured using a metabolic dye (XTT or TTC) or by viable counting after removing the biofilm from the surface of the microtitre plate. The XTT and TTC assays are time consuming and XTT is expensive. Different published studies showed that the CV staining is commonly used for measuring bacterial biofilms and the XTT assay is commonly used for measuring fungal biofilms (Li *et al.*, 2003). The result of the CV assay showed similarity with other published studies (Peters *et al.*, 2008). Compared to the XTT and TTC-reduction assays, the CV staining method was simple, cheaper and faster. In conclusion, the most appropriate susceptibility testing technique for *S. aureus* and *S. epidermidis* was the CV assay.

References

1. Donlan, R.M. (2001) Biofilms and device-associated infections. *Emerging Infectious Diseases*. 7(2): 277-281.
2. Gilbert, P., Das, J., Foley, I. (1997) Biofilms' susceptibility to antimicrobials. *Advances in Dental Research*. 11: 160-167.
3. Juda, M., Paprota, K., Jaloza, D., Malm, A., Rybojad, P., Gozdziuk, K. (2008) EDTA as a potential agent preventing formation of *S. epidermidis* biofilm on polychloride vinyl biomaterials. *Annals of Agricultural and Environmental Medicine*. 15: 237-241
4. Li, X., Yan, Z., Xu, J. (2003) Quantitative variation of biofilms among strains in natural populations of *Candida albicans*. *Microbiology*. 149: 353-362.
5. Peters, E., Nelis, H.J., Coenye, T. (2008) Comparison of multiple methods for quantification of microbial biofilms grown in microtitre plates. *Journal of Microbiological Methods*. 72: 157-165.
6. Thein, Z.M., Samaranayake, Y.H., Samaranayake, L.P. (2007) In vitro biofilm formation of *Candida albicans* and non-*albicans* Candida species under dynamic and anaerobic conditions. *Archives of Oral Biology*. 52: 761-767.

Figure 6.2　An example of a poster

AUTOIMMUNE HAEMOLYTIC ANAEMIA (AIHA)

INTRODUCTION

Autoimmune haemolytic anaemia (AIHA) is a type of haemolytic anaemia which causes destruction (haemolysis) of the body's own erythrocytes. There are two main types of AIHA:

- Warm AIHA: the auto antibodies attach to the erythrocytes and destroy them at temperatures equal to normal body temperatures.
- Cold AIHA: the auto antibodies attacks and destroys the erythrocytes at temperatures below normal body temperature.

CAUSES

In approximately 50% of cases, the cause of AIHA cannot be determined. But it is known that this can be caused by other diseases such as systemic lupus erythematosus (SLE) or due to certain drugs such as penicillin, cephalosporins etc., which are absorbed from the cell surface.

Genetic causes of AIHA are:

- erythrocyte cell membrane abnormalities e.g. spherocytosis;
- haemoglobin abnormalities e.g. thalassaemia;
- enzyme defects e.g. glucose-6-phosphate dehydrogenase deficiency.

PATHOGENESIS

There are three mechanisms that are activated as a consequence of the erythrocyte-antibody interaction:

- complement –lysis;
- adherence of C3-coated erythrocyte to complement receptors on macrophages;
- adherence of IgG-coated erythrocytes to Fc receptors on macrophages.

EPIDEMIOLOGY

- This type of anaemia affects women more than men.It mostly presents in middle-aged and older individuals.
- People of West African origin are mostly affected by this, due to glucose-6-phosphate dehydrogenase deficiency.
- This type of anaemia may also occur in individuals from Southern Europe, the Middle East and the Indian Subcontinent.
- The prevalence of this disorder is less common in children and adolescents than in adults.

SIGNS

- General pallor
- Pale conjunctivae
- Hypotension if severe
- Mild jaundice
- Leg ulcers
- Right upper abdominal quadrant tenderness
- Bleeding

SYMPTOMS

The most common symptoms are:

- Tachycardia
- Angina
- Weakness
- Gall stones, can develop in patients with persistent haemolysis causing abdominal pain
- Haemoglobunuria (dark coloured urine)

REFERENCES

A.V. Hoffbrand et al. (2001) Essential Haematology. 4th ed. London: Blackwell Science Ltd.
Martin R. Howard & Peter J. Hamilton (2002) Haematology. 2nd ed. London: Harcourt Publishers Limited.
http://www.patient.co.uk/showdoc40091014
http://emedicine.medscape.com/article/201066-treatment
http://bestpractice.bmj.com/best-practice/monograph/98/basics/epidemiology.html

Autoimmune Haemolytic Anaemia

DIAGNOSIS

- Direct antiglobulin Coomb's test performed. Here, patient's plasma is mixed with normal erythrocytes to determine free antibodies in the plasma. If agglutination occurs, indicates a positive result.
- The test is ≤ 98% sensitive, false-negative and occurs due to very low density antibodies or if the autoantibodies are IgA or IgM. The gel card direct Coomb's test is thought to be higher in sensitivity and specificity.
- There are three patterns of the direct antiglobulin reaction:

1. Positive reaction with anti-IgG and negative with anti-C3. This is common in idiopathic AIHA and in the drug or α- methyldopa.

2. Positive reaction with anti-IgG and anti-C3. This is common in cases with SLE and idiopathic AIHA, (usually warm antibody haemolytic anaemia).

3.Positive with anti-C3 but negative with anti-IgG. This occurs in idiopathic AIHA, usually warm antibody haemolytic anaemia, when the IgG antibody is of low affinity.

- AIHA can be categorized into two according to optimal temperature of antibody activity.

1. Warm-reacting autoantibodies (usually IgG), which are optimal around 37ºC

2.Cold-reacting autoantibodies (usually IgM), which are optimal at 4ºC

TREATMENT AND MANAGEMENT

There are many different methods of treatment and management:

- Folic acid is given as autoimmune haemolytic anaemia.
- Penicillin and ampicillin
- Transfusions should not be carried out unless necessary as in AIHA type matching and cross matching is difficult.
- Splenectomy.
- Iron therapy is used when a patient has severe intravascular haemolysis
- Corticosteroids are used as agents which have anti-inflammatory properties. They are usually used for the warm type of AIHA.

Cold types should be managed by keeping extremities warm

A group of student names goes here

Figure 6.3 A group poster displaying information about a case study

Further reading

Cornell Centre for Materials Research (no date). *Scientific Poster Design* [online]. Available from: http://www.cns.cornell.edu/documents/ScientificPosters.pdf [Accessed October 2009].

Divan, A. (2009) *Communication Skills for the Biosciences*. Oxford: Oxford University Press.

Hess. G., Tosney, K. and Liegal, L. (no date). *Creating Effective Poster Presentations: An Effective Poster* [online]. Available from: http://www.ncsu.edu/project/posters/NewSite/index.html [Accessed October 2009].

7

Oral Presentations

About this chapter

This chapter will look at the occasions when you are asked to give an oral presentation. We will look at ways in which you can structure your presentation, and give tips on engaging your audience and on what you should avoid. We will also look at the use of PowerPoint in presentations and how to control those nerves!

Introduction

Oral presentations are required in academic and professional life as an effective method of communication. A job interview often will include a short presentation. A manager in the workplace may need to communicate information, such as a training protocol, to his or her staff. Scientific meetings and conferences use oral presentations as a method in which scientists inform their peer group about the findings and results of their research work. Politicians and would-be (or existing) prime ministers continue to use the 'party political' oral presentation to communicate to the public. A survey of employers in science and other industries has identified the ability to communicate, particularly orally, as the most important skill a graduate should possess. Undergraduate and postgraduate study programmes include oral communication as one of the transferable skills which students should develop. Many students shy away from this aspect of their study and may even avoid those modules that include oral presentations in their assessment. However, this is one area of your academic life in which

Communication Skills for Biosciences   Maureen M. Dawson, Brian A. Dawson and Joyce A. Overfield
Copyright © 2010, John Wiley & Sons Ltd.

constant practice undoubtedly brings improvement, confidence and self-esteem. In addition, your skills in organizing and presenting scientific information are enhanced.

Rules to follow and pitfalls to avoid

Have you ever sat in on a presentation that you considered really awful? Think about what made that presentation so bad. Was it:

- too slow?

- too fast?

- too long?

- boring?

- irrelevant?

- too quiet?

- too loud?

- incomprehensible?

- badly planned?

Your own experiences of attending poor presentations should be the best indicator of what not to do, although we appreciate that nerves can sometimes lead to the problems mentioned. Remember that the delivery of information in an oral presentation is controlled by you, the speaker. This is different from reading a textbook in which the reader has control, can speed up, slow down or re-read sentences. The individuals in your audience can absorb only a little information at a time and are required to use their listening skills whilst simultaneously reading from a projection screen. Therefore you should pace your presentation with care; it should be such that the audience can keep up, but not be so slow as to be laboured and boring. Practise beforehand and enlist a friend to listen and give advice. Using PowerPoint for the presentation allows you to set a specific time span within which the presentation will run. This can be a benefit in that you know you will not overrun, but unless you practise carefully you will find the presentation 'takes over' and the slides run on to the next before you are ready!

Overrunning or finishing too soon both indicate that you have not planned and practised. If your presentation is being assessed as part of your studies you are likely to lose marks. Make a list of the points you definitely want to make. This should include:

- introductory aspects/an overview of what is to come;
- the 'main body' of the information;
- concluding remarks/summary.

Some advice taken from the example of television broadcasts is to adopt the 'ten o'clock news' style, which adopts the following format:

- tell them what you are going to tell them;
- tell it to them;
- tell them what you have told them!

Finally, how will you know when it is time to stop? Some tips to keep track of time include:

- taking off your wristwatch and placing it on the desk or lectern where you can see it easily. This prevents looking at your wrist in an offputting and unprofessional manner;
- if you own a mobile phone, you could use it to signal an alarm when you have one minute remaining.

Note how long one minute lasts when you are rehearsing so that you know how much information you can include during that vital last minute. Jump to the concluding remarks so that you end smoothly. Do not try to 'fit it all in' by talking more quickly. Above all, do not overrun! Imagine you are speaking at an important international conference. The next speaker has flown 5000 kilometres to give his/her talk. It is near the end of the day's events – and you are still speaking! Perhaps all the earlier speakers overran by two minutes. The last speaker will not be able to do his/her presentation and will fly back home dissatisfied. The audience are dissatisfied, as are the organizers.

Timing may be easier if you use the features provided in PowerPoint. The presentation can be set to run for a specified time, for example, ten minutes. In this situation the slides change automatically after the specified time and the total presentation is completed in the set time. The use of this feature requires

practice and discipline to ensure that you finish what you have planned to say in the time allocated for each slide. If you decide to add extra information the slides will be out of synchrony with your prepared speech.

How to keep your audience interested

Some questions you should ask before you prepare your presentation are:

- **Who will your audience be?** Are you speaking to fellow students, your teachers or lecturers, researchers, the general public, prospective employers? Perhaps your audience is a mixture of all of these;

- **What is the presentation for?** For example, is your presentation aimed to improve understanding or to impart knowledge, or to be assessed on your own abilities? Is it for a job interview? Each of these situations will require you to focus the information in the most appropriate way.

You should maintain frequent eye contact with your audience. Try not to look only at the screen, or your notes, or the 'lecturer in charge'. Worst of all, do not turn around with your back to your audience, while you read from the screen – it is tempting to do this if you are nervous but is not recommended. A good exercise to make the audience feel included is to rehearse in a small group by getting everyone to stand at the start of your talk. They may sit down once you have gained eye contact with them. By the end of your talk each individual should be seated.

Your audience must be able to hear you and understand your words. Be aware of your own voice and the vocabulary that you use. Speak clearly, practise projecting your voice to the back of the room, remember to take a deep breath and keep breathing regularly, especially if you are nervous (more about nerves later in this chapter). Avoid the use of slang, colloquialisms and abbreviations, unless you first explain what they are.

If you are using slides in electronic media to present your information, aim to vary the content and limit the amount of information on each slide. This should be a set of key phrases, supported by tables, illustrations, figures and references if appropriate. You should not display the whole talk and you should be practised sufficiently so that you do not need to read from your slides. If you are using PowerPoint do try the function for 'speaker's notes' (Figure 7.1). This allows

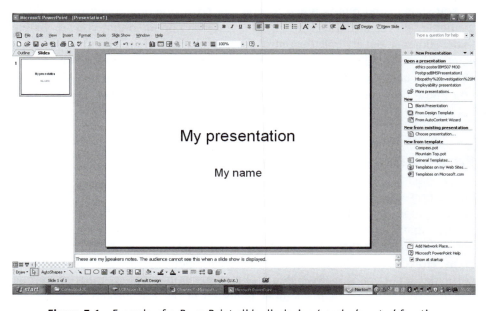

Figure 7.1   Example of a PowerPoint slide displaying 'speaker's notes' function

you to type your own words underneath the slide. When you are in 'Presenter' view, you can use these notes as a Prompt. However, you will need to have a computer with two monitors, one in slide show view which will be projected on the screen, the other in presenter view for you to read. Alternatively you can print out the speaker's notes and use those as your prompt.

How to deal with stress and nerves

Even the most experienced and laid-back of speakers, from politicians and news readers to lecturers and students, have times of nerves and stress. Remind yourself of this fact, take a deep breath and keep calm. Physical signs of nerves or stress include damp, sweaty palms, shaking hands, blotches and flushing of the skin, especially of the neck and throat area, and a faltering and stammering voice. These signs are often very obvious to you, the speaker, but may not be at all apparent to an audience. It may help to have something in your hands – a pen or pointer for example – and have a glass of water to sip. Printed notes on A4 paper are not recommended because you are likely to lose your place when looking up at the audience, and the paper shakes if you do! Small cards, written on one side only, and numbered in sequence in the top corner, are a good method for providing your own notes. One of the best ways to deal with nerves is to practise thoroughly beforehand. Try to anticipate any questions the audience may ask and get your answers prepared beforehand.

You should find out as much as you can about the venue and the facilities available beforehand. Check where you will stand or sit so that you are visible to the audience but are able to control any equipment such as a computer mouse, electronic whiteboard, visualizer, microphone, flip-chart, window blinds etc. Find out where the light switches are if you wish to dim the lights. Make sure you know how the audio-visual equipment works and whether any technical help will be available to you.

Presentations with a group of participants

Group presentations are good for developing your teamwork skills. Many university programmes will use group presentations during the first year, as this ensures that you will get to know people within your course. It also helps nervous students if they are presenting within a group and can share their concerns. However, presenting information in conjunction with your peers brings a particular set of concerns to students. How long should each member of a group talk for? What if one team member is absent at the last minute, or if a team member has not 'pulled his or her weight' in contributing to the planning of the presentation? It is also essential that a group presentation has been genuinely produced as a team and is not just a collection of separate PowerPoint slides that have never been seen by the whole group until the final presentation!

Some suggestions

The speakers. Identify who will be speaking, in what order, and for how long in each instance. Timing is particularly important and you should ensure that any one individual does not speak for longer than the allotted slot. Begin by introducing yourselves – it is useful to list the names and affiliations (a piece of information about each individual, such as the programme or module being studied, or where they are from). This may be on a PowerPoint slide or on a handout.

Do not take too long on the introduction to the topic as you may be short of time later. One team member should keep time and have a pre-arranged signal for moving to the next speaker in the group, for example, simply standing up (if sitting), or moving from one seated position to a seat nearer to the speaker, or raising a hand, using fingers to indicate the remaining length of time.

Assessment of presentations. If the presentation forms part of your programme of study, it may be assessed by a tutor and the marks or grades will be provided for feedback and may contribute to your coursework mark profile. Begin by finding out how the presentation will be assessed. Are marks split evenly between group members? Will marks be lost if one member has not contributed? Covering the topic adequately may be the most important aspect for assessment; thus it is necessary that another team member can step in if someone is absent at the last minute, or has not contributed in the planning stages.

Figures 7.2 and 7.3 give examples of how oral presentations might be assessed. The feedback sheet shown in Figure 7.2 uses a tick box to grade the presentation, scoring from 'poor' to 'excellent'. Key aspects of the presentation are assessed, including references/sources of information selected, aspects of the subject/content, presentation issues such as audibility and structure, the use of visual aids and finally the level of understanding regarding the subject matter of the presentation. The example in Figure 7.3 is specifically designed to assess a case study approach, often used in the biosciences, to evaluate aspects of a given disease or disorder.

Rehearsing your presentation. We recommend that you rehearse your presentation in front of an audience – it can be your flatmates, friends, the pet cat or even a mirror! This helps you to remember the information, gives an idea of length and identifies words that you do not understand. In this instance you should look up the meaning of a word in a dictionary or on the Internet. Be wary of using Wikipedia on the Internet, as this source is not necessarily accurate. If you come across long or complicated scientific or medical terms during

Feedback sheet for oral presentation	Poor	Significant improvement needed	Some improvement needed	Good	Excellent
Information sourced and referenced					
Appropriate texts/reviews papers sourced for the topic					
Accurately referenced					
Oral presentation					
Appropriate aspects included					
Appropriate balance/emphasis between aspects					
Subject content					
Relevance of information used					
Depth of information					
Quality of statements: information content; understanding conveyed; conciseness					
Presented clearly and effectively:					
Audibility					
Structure					
Cohesion					
Visual aids:					
Clear					
Readable					
Pertinent					
Understanding:					
Appreciation					
Awareness					
Perception					

Figure 7.2　Example of a feedback sheet for oral presentation

your background reading, you should find out the meaning, and then substitute your own words to explain them more clearly.

There may also be words you do not know how to pronounce and it does not give a good impression if you are stumbling over long words. Try to find out from a tutor or colleague how to pronounce new terminology, and practise saying it repeatedly well before the presentation. If, during your presentation,

Topic: _____ Evaluator: _____

	Exc.	Av.	Poor	Comments
Content				
Explanation of symptoms				
Differential diagnosis				
Well structured; organized; practised				
Correct diagnosis				
Rationale/evidence for diagnosis				
Enthusiastic, engaging; aware of the audience				
Appropriate support materials; integration into delivery				
Confident performance; voice control; appropriate gestures				
Time management				
Referencing				

Figure 7.3   Example of feedback sheet for case study/disorders presentation

you admit to your audience that you do not know how to pronounce a word, you are giving away the fact that you are not sufficiently prepared.

Using visual aids. Visual aids are just that: an aid to help the audience focus on aspects of the presentation. Therefore the visual aid should not contain too much information. It is usual to talk through or explain the information provided. This takes more time than you would expect and we strongly suggest that you practise talking through the information on the slide. Examples of visual aids include paper flip-charts, electronic interactive whiteboards, electronic visualizers, which magnify and project onto a screen whatever is placed beneath the lamp, and overhead projectors which require transparent acetate sheets. Photographic slides can be projected onto a screen. However, the most popular visual aid is Microsoft PowerPoint, which employs data projection. Whatever your choice of visual aid, do ensure that you are proficient in its use, or be certain that technical help is readily available. Be prepared by taking marker pens if you intend to use a flip-chart, or something to use as a pointer, for example, a laser pen, or, if using a PowerPoint presentation use the computer mouse to move the cursor arrow in order to point to specific areas of the slide for emphasis. It is important not to put too much information on each slide. Most of the information will be in your 'speaker's notes'.

Use of colour is recommended for added interest or emphasis. Choice of colours should be made with care as some colours are not clear when displayed and can make reading difficult. A dark background with pale letters can be very effective and is thought to be less tiring for the eyes of your audience. See Chapter 6 for more information on the use of colour. A background image may be added but do check that it is not too overpowering or fussy, detracting from the information on the slide.

Another tip is to reveal the information gradually in stages. This can help the audience to follow where you are on the slide, point by point, rather than overwhelming the eye with too much information. Again, PowerPoint has many features to aid this style of delivery, in particular the 'slide transition' feature. Information provided using acetate sheets or visualizers can be concealed and revealed in a similar way by covering the information with a blank sheet of paper, which is then gradually moved manually to reveal the information you wish to talk about, or by overlaying one acetate upon another, each building upon the information displayed.

Selection of the type and size of font is key to a good presentation. There are two aspects to consider in selecting font size. First of all the audience should be able to read the information clearly at the back of the room or venue. Secondly, varying the font size and type can be used to indicate which phrases are main headings and which are sub-text. It is advisable to use the same font size for headings of the same level of importance. We suggest trying a font size of 40 for main titles and 32 or 24 for subheadings and text.

The choice of font is also an important consideration. PowerPoint provides a massive choice, so where do you start? Fonts are split into roughly two groups – 'sans serif' and 'serif'. An example of a sans serif font is Arial, whilst Times New Roman is a serif font. As a guide, sans serif fonts are clear for the eye to read, and are useful in visual presentations. If you wish to be less formal, consider Comic Sans MS. This has the effect of making your presentation more 'accessible' to the audience. Figure 7.4 shows examples of the three fonts described here.

However, caution should be exercised when using PowerPoint. Although it provides a wide variety of features, it is not advisable to dress up the presentation with too many features, as they soon become 'gimmicky' and detract from the message you are trying to get across.

Use of images in PowerPoint presentations

The Internet provides an excellent selection of images which can be downloaded to provide illustrations of diagrams, photographs, pictures etc. This will

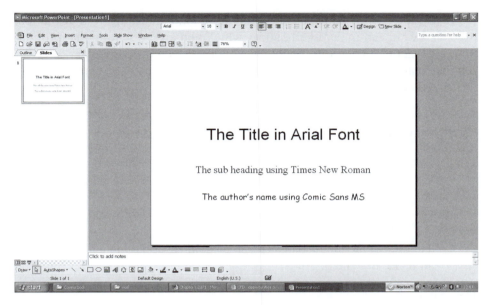

Figure 7.4   A PowerPoint slide illustrating three different fonts

make your presentation more interesting, in addition to using illustration as a mechanism to aid understanding. It is important to be aware of copyright: some images downloaded from the Internet are freely available, whilst others cannot be reproduced in full. Thus it is essential always to state the website and author(s) from which the image was taken. You should never cut and paste an image to be used as if it were your own work. This not only contravenes copyright laws but is also plagiarism. The same rule applies to images or illustrations photocopied from textbooks. You should cite the source on the specific slide, and provide the full reference at the end of the presentation with your other references.

A word about the content. You will probably use a 'précis' format, which means using short sentences, phrases or key words as a prompt on the overhead or PowerPoint slide. Do not expect the audience to read large amounts of text. In addition you should include tables, diagrams, figures or other illustrations. Scientific or other data must include units of measurement, for example: m, kg, s, A, K, mol (Chapter 4), and these should be SI units if you are addressing the scientific community. Any illustrations need to be explained to the audience. Talk through what they represent; take time to explain the horizontal and vertical axes on graphs. Although **you** know what a figure means, your audience needs explanation and time to work it out.

Finally, the PowerPoint slides may include a logo indicating your affiliation, by which we mean the umbrella under which you are presenting the information, for example, your school, college or university, or a particular society. A running header and footer may also be used to provide the date, names and title of the presentation on each slide.

Summary: a quick guide to oral presentations

Nerves

- Everyone has them to some extent.
- They are not usually apparent from a distance.
- Holding something, such as a pen or pointer, or having a glass of water to hand, may help.

Rehearsal

- This gives you an idea how long your talk will last.
- It helps to fix the topic firmly in your mind.
- Go through your slides to make sure you know what is coming.

Content

- Begin by introducing yourself.
- Introduce the topic.
- Display the title on a PowerPoint slide.

Use 'News at Ten' style

- **Introduction:** tell them what you are going to say.
- **Content:** give them your talk.
- **Summary:** tell them what you have told them.

Tables/figures

- Must have headings and include scientific units.
- Talk your audience through them, explaining any data and statistics, and labelling of axes for graphs or charts.

Slides

- Projection should be large enough for people at the back of the room to see.

- Lower case font is recommended: 40 point for titles, 32 or 24 point for text.

- Do not include too much information or long sentences (five points per slide is ideal).

- Bullet points and short phrases work best.

- Do not change slides too quickly! One slide per minute is a rough guide.

Ending

- Indicate to your audience that you have finished.

- Use a conclusion or summary slide.

- Thank your audience.

- Ask for questions.

Questions

- Allow time for dealing with questions.

- It is useful to repeat the question, perhaps in a shortened form.

- Make sure that the question has been heard by all.

- If you do not know the answer, just say so, though you could perhaps try to make suggestions; you could also ask if anyone else knows.

Finally, relax, take a deep breath and be yourself. Good luck!

Further reading

Fraser, J. and Cave, R. (2004) *Presenting in Biomedicine: 500 Tips for Success*. Oxford: Radcliffe Medical Press.

Jones, A.M. (2003) *The Use and Abuse of PowerPoint in Teaching and Learning in the Life Sciences* [online]. Available from: http://bio.ltsn.ac.uk/journal/vol2/beej-2-3.pdf.

8

Preparing a *Curriculum Vitae* and Job Application

About this chapter

At the beginning of your final year, you will probably start to think about getting a job or possibly applying for a further course. This is the time, if you have not already done so, to think about preparing a *curriculum vitae* (CV), which is a short summary of your education, qualifications and employment history. A CV can be kept electronically and regularly updated as appropriate. Some of you will undertake a period of industrial placement as part of your programme. If so, you will need to prepare your CV much earlier. This chapter gives specific guidance when preparing a job application, including the CV. It will also give guidance on preparation for, and conduct at, interview. The aim of this chapter is to give you help and guidance that will enhance your chances of getting the job, or winning a place on the course.

Introduction

Before you get round to applying for a job or course, you will obviously need to send a letter asking for further details of the post, and where necessary an application form. This letter needs to be both clear and brief. If you are applying for a job, you should identify clearly the title of the particular job together with any job reference, and where you saw the advertisement. For a course or research degree, you should ask for further details of the range of appropriate

Communication Skills for Biosciences   Maureen M. Dawson, Brian A. Dawson and Joyce A. Overfield
Copyright © 2010, John Wiley & Sons Ltd.

programmes available in the department or school to which you are applying. If, of course, you have already identified a specific programme, you should spell it out in your letter. Do not forget to include your current address and to give your name in typescript below your signature, in case your signature is indecipherable to the recipient. Do not go into details of your qualifications or employment track record at this stage. These will form part of the covering letter for your formal application once you have received the further details and forms. Very often in response to your enquiry you will be sent an information pack. In addition to the details of the job or course this may well include an application form, together with instructions for completion, as well as more general background about the company or department/school to which you have applied. Always read all the information sent to you. It will take time, but reading it will ensure that you will know what you need to give as part of your application, and help identify any queries that you have.

When you are applying for a job, whether placement or permanent, or for a further course in a university or college, you will be expected to complete an application form, supported in many cases by a CV, as well as some kind of covering note. In any case, you ought to have a formal CV tucked away on your computer. You will also be expected to provide the names of referees to support your application, and the choice of referees can be very important.

If you are applying for a placement during your course as part of the requirements for your degree/diploma, you will generally find that your tutor or supervisor will be able to give you advice, and there may also be departmental/ school guidelines. When you are reaching the end of your course, you will be looking around for jobs, or for further courses at graduate level. It is advisable to get in touch with your university's or college's Careers Service when you reach your final year or even earlier. Staff there will be able to give you independent and informal guidance on your options. They will help with applications for jobs, either through 'milk-round' interviews, or with guidance on applying to individual companies. They will also be able to give you advice on the choice of graduate courses, and where to apply, if you want to continue your studies at a more advanced level. When you are applying for a graduate course, it is always sensible to talk to your tutor or supervisor. He or she will have had lots of experience in helping students to apply, will, most likely, be one of your referees, and will consequently need a copy of your CV for background information.

You may very well discover when you see the Careers Service or receive details in connection with a job application that it uses its own equivalent of technical language. You will find some of the terms and a brief description below.

The *curriculum vitae*

As already stated, a CV is a document that gives an account of your previous employment (where appropriate), your education history and qualifications obtained, as well as details of other experience and skills. A CV is also commonly referred to in the United States and elsewhere as a resumé.

Remember, when you are applying for a job where there is no formal application form, you will also need to include some details in your CV or covering letter that may not normally be included: for example, information about your gender or marital status and, of course, a statement of why you are applying for the job. There will inevitably be differences in the sort of CV required for a company employer and that for a university teaching post. Universities will generally require information about publications, research grants and details of teaching and research activity. Bear in mind that a CV is a dynamic document that evolves as your career develops. The sort of CV you have when you are applying for your first job or placement will be very different from the one you will have when you apply for a post of professor of biology.

Before you start writing your first CV

You will need to have to hand a lot of key information, such as qualifications and dates obtained, as well as details and dates of jobs you have held. For your educational qualifications, it is wise to have copies of the original certificates. You will always find if relying on memory that you cannot recall all the actual details of your qualifications, and may miss some, unless you have a photographic memory and instant recall. Under no circumstances should you claim to have any qualifications that you do not. The company offering the job or the admissions tutor for the course will check up such details, or will compare what you say with what your referees say. If you have falsified details, like claiming that you have a first-class honours degree when you do not, your application is likely to be disqualified. If the falsification has been discovered late in the process, a job or course offer will be withdrawn.

The introduction to your CV

The introduction to your CV should always contain personal details, including full name, home address, telephone number and email address. Always give all the names that appear on your birth certificate or passport. This will avoid problems later, for example, in connection with National Insurance details or information about your permission to stay either as a student or worker status.

Avoid using university/college address information and telephone numbers, such as those for a hall of residence, and a university/college email address. These change much more frequently than home contact details. You will often find that employers or universities will include in their application forms opportunities to list your current address for correspondence and telephone number, and you can give your university/college address here and your mobile number if need be. Your covering letter should also normally include your correspondence address. If you know that the application process is likely to be prolonged, as it often is for admission to courses, give your home address as well in your covering letter, unless, of course, you are an overseas student likely to stay at your current address for some time. It is not normal nowadays to give gender, marital status, date of birth or nationality, because of the dangers of discrimination on the grounds of, say, age, gender or race. If there is no application form, and this information is requested, adapt your CV as necessary. In a short CV, after the personal details, you can include a brief statement of the reasons you think that you should be considered for the job.

Previous employment and education

The order in which you list your previous employment and education can be varied.

- In university/college applications the emphasis will be on educational qualifications, and they will come first. A detailed list of qualifications is usually required. Put dates and institutions attended before qualifications obtained. In both cases you would normally put them in chronological order, starting with the earliest.

- For job applications in industry, the emphasis is usually on previous employment history, and that will come first. Educational qualifications are needed but employers in industry would not normally expect to see details of GCSEs, unless they wanted a specific qualification, such as a grade in mathematics. Just give numbers and dates obtained and a summary of grades. With secondary-level qualifications, if you are applying before you have your degree, or are applying for a placement, it is sensible to give full details of awarding board, dates and grades.

- For previous employment, it is normal to give details of your most recent post and current salary first, followed by a summary of other posts and dates in reverse chronological order. You may wish to give more details about your current post, and aspects of your earlier employment, particularly where they are relevant to the job application. If you are applying for your first job or placement, then give some details of summer jobs or voluntary work where this is relevant.

- Where appropriate you should give details of professional training or qualifications, such as IT awards or language skills, for example, fluency in French.

- In addition, give details of voluntary/community work where it is relevant to your current application.

Publications and research grants

Universities and colleges will always ask for details of publications and research grants that you have been awarded when you are applying for an academic or research job. It is not expected that you would have publications if applying for a research job straight after graduation. You should be realistic in making applications: a high-level research post is unlikely to be given to a new graduate. However, a research assistant post, particularly if related to your undergraduate project, is within your grasp.

Lists of publications, if you have them, for example, as a result of a postgraduate research degree, such as a PhD or an MSc, should be incorporated into your full CV, but be prepared to adapt your CV as appropriate. Universities will often give guidance on the order in which publications should be listed, with priority given to research monographs first, followed by journal papers, down to newspaper articles and the like. Within each group they should be listed in chronological order. Often they will also expect grants from research councils to be listed before industrial grants. Again, they should be listed in chronological order within each group.

By contrast, when you are applying for a post in industry, unless it is specifically research-oriented, the employer will not generally want to know about publications. Nevertheless, if your publications or grants are relevant to your job application, refer to them in the statement that you include to emphasize your suitability for the job. You would normally include this after personal details in a short CV or in the covering letter.

Final part

It is sensible to give some information about your hobbies or any community work that you are involved in. This may be particularly important where you have won awards, such as a sports medal, or perhaps a commendation for your volunteering work. If you are, for example, captain of a sports team, this will indicate to your potential employer that you have leadership qualities, or, in the case of a course application, that your interests extend well beyond your current course.

Normally you will be expected to include a list of referees at the end of your CV. The number required for a particular application will vary, but always make sure that they are current, and that you have sought their permission. For your first job include your tutor or supervisor and your head of department/school, or someone from outside with recognized standing who can give you a good (and accurate) reference.

General points

You should always bear in mind the following when preparing your CV:

- lay it out neatly and carefully, generally in two columns, with titles (for example, **full names**) to the left in bold, and details to the right in normal font;

- there are companies that provide model CVs. It is very important if you use one of these that you do not cut and paste indiscriminately, for example, a model statement about background that you think is particularly suitable for your application. Some universities and employers have started to use plagiarism software to identify where an applicant has copied whole passages from models;

- update your CV regularly;

- always have a full CV as your master copy and keep this up to date. You can then adapt it as the need arises. Some employers, for example, will ask for a brief CV. In this case aim for one or two pages. Others may ask for a full CV and possibly a complete list of publications and grants;

- be prepared to trim and tailor your master CV to produce an effective alternative version that matches the requirements of the job or course for which you are applying. Be very careful when adapting a CV that the adaptation refers to the current job, and not one you have previously applied for (this does happen!).

Covering letters

A covering letter is intended to let the person who receives it know what you are applying for. It will also allow you to set out your reasons for applying for the job or course, and to emphasize key points that you think will support your application. This will give your potential employer or admissions officer an impression of your suitability and of your ability to identify and set out simply what you see as the key features of the job or course. In addition, a well-written letter can make the difference between your being shortlisted, and your application just being left among the bundle of applications that are not really considered seriously.

Before you start

Read the advertisement and further particulars for the job or the course handbook very carefully, as well as any other information that you have been given in the application pack. In some ways this is like reading the question and initial background material for an essay, or proposal for a research project, except that the information that you have been given in the pack may well be considerably longer, and contain much more detail. The pack will give you the pointers for tailoring your covering letter, as well as any personal statement you will need to include in your CV.

When you are applying for a job, you will obviously need to think hard about your reasons for applying for it. Whatever the economic climate, it pays to set out clearly why you are applying for the job and your career aspirations. It is also important to focus on what you consider to be your most appropriate experience and qualifications for the job, and to bring them out very clearly in the covering letter. Try to get some up-to-date information about the company or institution to which you are applying. Read any literature that they send you. If shortlisted for interview you will be expected to have read it! Try to include something relevant in your covering letter to show that you are aware of, and interested in, the strategic aims of the company. It is also important to know if a company is under pressure, for example, where it has cut back on graduate trainee schemes. You need to look at this sort of thing, if only to consider the implications for your career in the company, and whether you would have a real future there.

If you are interested in pursuing another university course, it is clearly sensible to know about the general perception of the quality of the course and of the department/school that runs it, particularly its teaching standards and research quality. Where these standards have been rated highly, perhaps by external audit, the department/school will not hesitate to publish details in its course handbook or on its website. This can be particularly relevant when you are applying for a research degree. If you are applying for a taught course, you should also look in detail at the course content. It will not help your career if the course you undertake is only marginally related to what you see as the main direction of your future career. Equally, the relevance of teaching standards is important: you do not want to find yourself on a course where the level of support is significantly below what you would expect.

How to compose the covering letter

Your letter should be concise and focused. Remember that it will be supported by a CV, and possibly by an application form. You should set out precisely what job or course you are applying for. Your aim should be to spell out clearly why you are applying for it and why you think you are a suitable candidate. Always make sure that what you write is relevant to your application. Take care that your letter is well laid out and follows a structured sequence. It is clearly sensible to follow the order of the key points identified in the advert or prospectus. Do not be too informal in the language that you use, even if you know the person to whom you are writing, as your application is likely to be considered by others on an interview or selection panel. Always run a spell-checker and look carefully at your grammar. As has been implied earlier, a badly written letter will put off an employer or admissions officer; this is particularly so where the letter is full of spelling and grammatical errors. If you do have a disability, like dyslexia, or eyesight problems, get some support and, where appropriate, indicate the nature of your disability.

Layout

When you are applying for a job, always head your letter with the title of the job and the job reference number, where one is given. Most job application letters will be addressed to the human resources section of the company or institution, and they need to be able to identify the post, and who deals with it initially. You should then start your letter with a formal sentence along the lines of: 'I am applying for the post of..., advertised recently in ...'.

Give a code number for the job if there is one. You should then say why you are applying for the job. If you are applying for a job or placement, you will need to say what course you are currently studying and where and, if appropriate, your actual or expected degree result. You should then summarize what you see as your experience and strengths that are relevant to the post, following the order of the key points in the advertisement and person/job specification. You should end with a sentence or paragraph on how you see your role in the job as fitting into the company's strategic aims and how you feel that you can contribute to them.

Where applying for a course or research degree, you should spell out in the header the course or research programme you are applying for. Normally your letter will be addressed to an admissions officer in the department/school, or sometimes in the case of a research degree to a named member of staff. Some universities and colleges handle applications for graduate programmes centrally or at a faculty level, so it is important in this case to identify the course and department/school that offers it. As with a job application, you should then open your letter formally along the lines of: 'I wish to apply for a place on ...', followed by the name of the course and department/school, if necessary and, where appropriate, the year that you wish to start the programme. If you are applying for more than one programme, for example, where there are masters options and diploma options, you should name both, but state your preference for one. However, by the time you formally apply for a place, you should have identified the topic of the course you want to take. With a research degree application you would normally give the name of the degree, and the research topic that you wish to follow. Sometimes research studentships are targeted at specific research; in other cases they may be open departmental or faculty awards that do not specify a topic. In all cases you should identify the reasons for your interest in the course or research degree and emphasize why you think you are suitable, for example, by outlining particular strengths, such as a very successful undergraduate project. You should give details of the course that you are currently following and the name of the institution, and give either your actual or your expected degree results. If you are applying for a research degree, you should spell out in your final paragraph how you believe you can contribute to the research undertaken in the department or school.

Whether applying for a job or a course, it is always helpful if you set out your reasons for applying, details of your experience and why you believe you are a suitable candidate as bullet points. This can bring added clarity and focus to your application.

After you have drafted your letter, re-read it to check for errors and omissions and to ensure that it actually says what you want to say. If you have to submit a handwritten application (it does occasionally happen!), make sure that it is neat and legible. You should use a fountain pen or a good quality roller ball pen, rather than a biro.

In all cases you should give your current address; you should also give your name in typescript at the end of the letter under your signature. Do not forget to sign and date the letter.

Application forms

Very often, particularly with universities or colleges, you will be expected to fill a formal application form. This is to ensure that the information given by all candidates follows a consistent pattern. Where there is an application form, always complete it: the notes of guidance often say that if you do not, your application will not be considered. Always complete all sections of the form fully. Do not enter 'see CV' as the only entry for a section.

Application forms for a job will generally be different from those for courses, though there will be some similarities. Both will include detailed instructions on how the form is to be completed, and you must read these very carefully. The detailed instructions and the form itself will emphasize that all sections have to be completed. They will also state that a CV and other information provided will only be considered alongside the application form.

Application forms should, where possible, be typed or handwritten in black ink. Many application forms are now available as downloads from the website, which makes typing easier.

They will also include sections that are to be completed by the office responsible for handling them, for example, application number and date received. Do not write anything in these sections.

Job application forms

The form will normally start with a section that asks you to identify the post applied for and job reference number and where you saw the job advertised.

Forms will normally follow the sequence of the usual format for CVs; that is, an introduction section requiring personal details, followed by sections seeking details of education and employment history, together with a request for the names of referees. In the personal details section, you may be asked for details of your nationality. If you are not from the European Economic Area (EEA) or Switzerland, the appointment will be made subject to a successful application to work in the UK. The employer will be able to give you guidance on the procedures involved.

All forms will include a section that asks for additional or supplementary information. You should use this section to summarize the main and relevant points of your CV and to set out how your experience meets the criteria outlined in the job/person specification (see below). You should arrange these points in the same order as those of the job/person specification, always bearing in mind that the criteria for each are normally listed separately in the accompanying information pack. Where necessary, you should continue on a separate sheet. You should always summarize the relevant key points of your CV and indicate 'for additional information/details see CV', where appropriate. Do not give a detailed re-run of your CV in this section, unless the further particulars specifically say that you should not enclose a separate CV. Always make sure that what you include in this section is consistent with what you say in your CV and in your covering letter.

The forms will normally conclude with a section asking for the names of referees. Ensure that the names that you give are consistent with those listed in your CV, and that you have asked their permission to act in relation to this specific job. Normally the employer will send copies of the job particulars to each referee. However, always make sure that each of your referees has an up-to-date version of your CV.

The associated job/person specification will set out the job description and grade and will detail the formal responsibilities of the job and to whom the appointee is responsible. It will go on to describe the principal duties of the post. The person specification will list the skills, experience and knowledge required: for example, experience of supervising others, communication skills and, for example, a working knowledge of Microsoft Office. For both you should summarize your relevant experience, following the order of the points listed, in the additional information section.

You should follow very carefully the guidance notes that accompany the form. These will contain a definition of some of the terms used, like job specification. They will remind you to read the job particulars and person specification very carefully. They will tell you how to fill in the form, and generally warn you that a CV alone will not be accepted as an application. They will also give you information about other matters, such as your obligation to disclose information under the Rehabilitation of Offenders Act of 1974, and any Criminal Record Bureau (CRB) checks that the employer may be obliged to carry out, as well as details of any occupational health requirements. As with 'small print' elsewhere, it will point out that the employer will not accept responsibility for forms arriving late or being lost in the post or over the Internet. They will also inform you about the employer's policy on equality and diversity and the like. Sometimes you will also be required to fill out an equal opportunities monitoring form.

Graduate course application forms

As with a job application form, you will be asked for personal details; this will include information about fee status and any details about permission to study in the UK. The form will ask for information about courses for which you are applying in order of preference. It will also have sections on qualifications and work experience. As with the job application form there will be a section that asks for supplementary information, which you should use to provide information about the relevance of your educational or work experience. Where you are applying for a research degree, you will be asked to give an outline of your proposed research. Again, if need be, you should continue on additional sheets. In contrast with a job application, you will not normally be asked to provide a CV. Make sure that what you say is consistent with your covering letter.

There will be supporting notes that you should read very carefully. These will include a mix of detailed information about how to complete the form and 'small print' information. The key things here are likely to be information about fee payment as well as a statement that when you start your course, you are entering into a legal contact with the institution. If you are not from the European Economic Area (EEA) or Switzerland, the offer may be made subject to a successful application to study in the UK. The university that offers the course will be able to give you guidance on the procedures involved.

How to write a good personal statement

The personal statement, whether in the covering letter or in your CV, allows you to sell yourself. Both employers and admissions tutors will not only be trying to identify those who meet their requirements, but will also be looking for that 'special' factor. It is therefore sensible to show how you meet their requirements. More important is to set out what you see as the 'special' factors.

Where you are applying for a job, say clearly why you want the job and how you see this as developing and building on your career. You should back this up with a statement that includes:

- why you have chosen your particular career;

- what your future plans are and how this job will help develop your career;

- how your previous experience fits in with the job, and, particularly, how relevant it is;

- what you see as your strengths, such as the ability to manage people, or in your first job, the ability to get on well with people;

- what your personal skills are, and the things that demonstrate your skills, such as the Duke of Edinburgh Award;

- how you see your role in the job as fitting into the company's strategic aims and the way in which you believe you can contribute to them. If you are applying for an academic or research post, how and what you can contribute to the department's/school's or research group's aims.

For your course application you should provide supporting information similar to that in a job application, in particular:

- why you are interested in the subject;

- what specifically interests you in the course, or why you wish to undertake a particular project for a research degree;

- how you see the graduate or professional qualification helping your career development;

- how any previous work experience fits in;
- why you think you will be a successful student, for example, by demonstrating your ability to manage time to meet project deadlines;
- what other activities demonstrate your wider abilities, such as involvement in music or drama, particularly where you have been a lead performer, or in sport where you have been a team captain;
- if you are applying for a research degree, how you believe you can contribute to the research in the department or school and how you will get on well with the rest of the research team.

Interviews

This section will describe the run up to the interview and the interview process itself. It will give some tips about how to prepare for the interview, the sort of questions you might be asked, and how you might answer them.

Sorting out arrangements for the interview

If you have been shortlisted for consideration for a job or place on a course, you will get a letter from the prospective employer or from the admissions tutor inviting you to come for interview. The letter will give details of the date and time of the interview, and either the venue or the office that you should report to when you arrive. The first and most obvious thing to do is to check that you are actually available on that day and at the time set for the interview. If you do have a clear day, or couple of days to take account of travel times, let the person who sent the letter know immediately. If you have classes, tutorials or laboratory classes on the day of the interview or on the day that you need to travel, let everyone concerned know, and check that they are happy to let you miss your class. If there is a genuine and insurmountable problem with the date and time, immediately contact the person who wrote the letter: you may be able to reschedule your interview. If you do have a very good reason, such as a family funeral or a pre-booked holiday, and the employer or admissions tutor says that the interview cannot be rescheduled, think hard about it. Would you want to be employed by or follow a course run by an employer or department that is so insensitive and inflexible?

Never ask for rescheduling for flippant reasons. If it is a job interview a good employer will be willing to reschedule if you do have some long-standing commitment, particularly if they see you as a good prospect. Admissions tutors generally are a lot more flexible; they do not necessarily have to schedule, say, five interviews for one day. You could always explore the possibility of a telephone interview, or even an online interview, if, for example, you are at a university abroad. Universities and colleges in particular often have the facilities to do this. If you are applying for a place on a course in a university overseas, this is the sort of approach you need to take. Obviously such an interview will be a little more difficult, but it is worth a try. If you have applied for a job abroad, then the employer will want to talk to you about possible arrangements for travel.

When you have sorted out a date, write formally to accept the offer of interview. Always respond directly to the person named in the letter, and not just to some anonymous department or office. The individual may be a departmental secretary, or a member of the human resources staff of the employer. They will have responsibility for organizing the arrangements for the interviews.

Make a note of the date and time, and check out how you will travel to the venue. If it is a long distance, and the interview is scheduled for the early part of the day, you are likely to have to travel on the previous day and stay overnight. Equally, it may be a late interview, and you may have to stay overnight afterwards. If so, ask the person who sent you the letter if they are able to arrange accommodation, or recommend a suitable hotel. Sometimes they will have arrangements with a local hotel or hall of residence. They will, of course, generally be helpful, especially if the place of the interview is a large city. However, it is sensible to recognize that if you are travelling from Aberdeen to Plymouth, or Aberystwyth to Canterbury, you may have to set aside two or three days for travel and the interview.

Check if you are going to receive travel and accommodation expenses; this may influence how you will want to get to the interview. Public transport is normally the best option, unless, of course, there is no station close by, and the coach schedule is slow and infrequent. If you travel by public transport, you are likely to be far more relaxed than if, say, you have driven 200 miles on the motorway system followed by ten miles driving to get to the centre of a city. Check out the best train and public transport options, allowing plenty of time to get from the station to the venue. Book trains as far in advance as possible: this will save money. However, do not book an off-peak train journey just because it is cheaper, if it means that you have only twenty minutes to get from the station to your venue. If you have to book your own accommodation, go for a relatively inexpensive chain hotel and keep any receipts. It may also be worth checking if any friends living locally could put you up. It may be easier for you to drive a couple of hundred miles on the previous day, and stay with friends, with just a short journey on the day of the interview.

Preparing for the interview

The first thing to think about is what you will do if you are successful, and are offered the job or a place on the course immediately after the interview. Do you have other interviews arranged? Is the job or course your top choice? Work out a strategy to handle this. Employers will allow some time for you to consider accepting the offer, but not too long. Admissions tutors are likely to be more flexible: they know that students will have more than one application for a place.

What is the format of the interview?

Find out about the format of the interview. It may be a relatively informal interview with an admissions tutor or potential research supervisor. Equally it may be a very formal interview by a committee consisting of a number of staff; this is much more likely to be the case with a job interview. Read the information you have been given about the interview. Check out if you are expected to give a presentation, which is quite common these days. Find out what the topic and length of the presentation will be, and what format it will take. It may be on the basis of a paper presentation with questions or an oral presentation with a flip-chart. It is much more likely to take the form of an oral presentation using a program like PowerPoint to provide accompanying slides. If so, check out both the software and the hardware to be used. Save your presentation on disk or pen-drive, and if necessary be prepared to take your own laptop, as you may be more confident with it. Also, take a hard copy of your presentation just in case the computer crashes (it is not unknown!). Do not leave the preparation of your presentation to the last minute, and rehearse it several times in advance so that you are fluent in what you want to say, and confident in speaking while handling a computer. Draw on the experience you have gained from presentations on your current course. If you are using a program like PowerPoint, do not repeat word for word what is on the slide show; amplify the points made in your own words (see Chapter 7).

For technician jobs you may be required to take a test to demonstrate your competence in data handling or calculation. For many other jobs you may be required to undergo psychometric testing. Psychometric tests fall into two categories. Aptitude tests look at skills like verbal and numerical reasoning or other specific skills. Personality tests look at personality traits, such as tendencies to be obsessive, as well as the strengths and weaknesses of candidates.

The run up to the interview

A few days before the interview consider having a haircut or hairdo at your local barbers or hairdressers. Unkempt, straggly or scraggy hair can be a turn-off for some interviewers.

On the day/evening before your interview re-read the job and person specification or course prospectus to re-familiarize yourself with the details and to identify any questions that you want to ask. Also print out and re-read a couple of times the application that you sent in, and highlight the bits that you feel are important, like your strengths and experience. Think about the sort of questions that you may be asked and how you would reply to them.

If you are travelling on the day before, ensure that you have packed sufficient clothes and your washing kit, and that you have hard copies of your application documents and the applications pack that was sent to you, as well as any receipts in connection with travel expenses. Also ensure that your laptop, if you are taking it, is well charged, and that you have cables that will enable you to connect with other hardware.

On the day of the interview make sure that you are clean and well turned out. Wear clothes that are smart. If you have to travel on the day of the interview, try to refresh yourself before your interview. Even on the most modern high-speed trains you can get sweaty and grubby on a long journey. Avoid the temptation to drink alcohol or have a large meal before your interview. If you are travelling by public transport, make sure that you have plenty of time to get from the station to the venue. If there are transport problems that delay you, ring through to the contact number on your interview letter to let people know. You will inevitably feel nervous. Try practising your favourite relaxation techniques (or learn some, if you do not have any).

When you arrive at the employer's building or the university, ask at the general reception desk for directions to the venue, and the location of other places such as the canteen and toilets. Then go to the office that sent out the invitation letter and ask for the person who sent the letter. He or she will normally be responsible for organizing the interview, and will be able to brief you about the interview procedures. He or she will ask you about your claim for expenses; these must be backed by receipts or ticket copies. You will then either be taken to the interview room or a waiting room close by, where there may be other candidates waiting to be interviewed. Sometimes before a job interview candidates will be invited to a general session or lunch, largely so that the prospective employer can see how each individual interacts with other people. Be relaxed and polite, and do not show off!

The Interview

1. The interview should be a two-way process: it should allow the interviewers to follow up and explore information that you have given to help them make up their minds; it should also allow you to showcase your strengths. In addition, it will give you the opportunity to follow up further information about the job or the course.

2. The interview should be structured. In selecting its shortlist for a job or a place on a course the committee or group of individuals interviewing the candidates should always have a list of essential and desirable qualities. There will also be a chairperson or 'chair' appointed to run the interview process. Where the committee is interviewing more than one candidate, a good chair will have agreed with his or her colleagues what the main questions should be, and there is an expectation that each candidate will be asked broadly the same questions. Unfortunately there still remain some chairs or interviewers who like to play things by ear, and this can be unfair to candidates. All interviews have to be guided by policies for equality and diversity, and should be conducted on the basis of ability without any kind of discrimination on age, gender or the like.

3. As far as interviews for places on courses are concerned, particularly where there are many candidates to be seen over a period of time, the admissions tutor may be the sole interviewer. The same will hold with interviews for research degrees, where the project supervisor may conduct the interviews alone.

4. Listen to the questions. If you do not understand a question, ask for it to be repeated or clarified. There are still some interviewers who will

ask trick questions, though this is now less common. There are also interviewers who adopt an aggressive style: this may very well be the case where you are applying for a job in commerce or finance, and where the interviewer is trying to make a big impression with his or her colleagues. Do not be fazed, and answer politely. Nonetheless, if you do feel that the questioning is too aggressive, you may want to raise it with the chair, preferably during the course of the interview or afterwards in private, especially if you believe that you are being unfairly treated. Sometimes an interviewer will ask an inappropriate question, like what if you become pregnant, or have problems if children are ill. A good chair, or if it is a job interview, the representative from human resources, will stop such questioning in its tracks, as it constitutes discrimination. If they do not, tell the chair firmly that you do not feel that the line of questioning is appropriate, and make a note of your protest to take it up later if necessary.

5. Always make sure that your answers to questions are consistent with the details you included in your written application, and if necessary explain any differences, for example, where there have been developments between the application and the interview. You may find it helpful to have a hard copy of your application to hand to check the details. There is always the possibility that the questioner has either misunderstood what you have written, or confused you with another candidate.

6. If you are giving a presentation, stick to the time allotted, and do not overrun. Do not simply repeat the wording of your slide show; amplify and explain the bullet points. Be crisp and assured.

7. Ask any question that you have prepared in advance or, as appropriate, things that have arisen during the course of the interview when the opportunity arises. The chair should offer you the opportunity to make points about things that you feel have been overlooked. Identify one or two key issues, and set them out concisely and clearly. This often takes place towards the end of the interview. Do not go overboard and re-run the points that you made in your covering letter or CV.

8. At the end of the meeting thank everyone. You should also leave contact details, especially your mobile telephone number. The chair or secretary may want to contact you as soon as all the interviews have taken place and the committee has reached its decision.

Typical questions and how you might answer them

This book has provided many pointers to help you develop skills in communication. These skills will help you to perform successfully in an interview for a first job, and especially for a place on an advanced course or research degree. What you have learned during your course should also have allowed you to identify the skills and experience that you will need in employment or for a further advanced qualification, in particular:

- working to deadlines;

- communicating with others effectively;

- working with others in a team;

- researching and selecting relevant information;

- problem solving

Your final-year project is an excellent way to showcase many key skills to a potential employer. Make sure you know your project in some detail, and be prepared to draw on it to highlight examples of the skills you have acquired. Take a copy with you to show the interview panel. Even if they only see the title and cover, this can leave a visual impression. You could also take an A4-sized print of a poster you have completed.

During the interview you are likely to be asked to give examples of your skills and experience, such as:

- **Working independently.** Start your answer along the following lines: 'When working on my project I had to …'. You should give examples of how you conducted your COSHH evaluation, thus showing knowledge of health and safety, as well as independent information retrieval. You may have worked individually in the laboratory for long hours. You will have planned your timetable to complete the project, including key stages that had to be completed within the project. You could also cite any occasions in paid employment outside your course when you have had to work independently.

- **Working in a team.** Take as an example how you prepared a poster with your colleagues and spell out how you had to meet regularly, email each other or plan the tasks for each individual in order to complete the job by the deadline for presentations.

- **Effective communication with others.** A good example is how you prepared and presented many oral seminars and your familiarity with PowerPoint presentations. Equally, you could give examples of other experiences in your degree in which you were required to communicate orally or in a written form.

- **Knowledge of scientific information.** Start your answer with something like: 'During my degree I studied ...'. Make sure that you are absolutely familiar with the topics that you have studied during your degree, and how they might be applicable to the job or course you have applied for.

- **How you would apply your knowledge and skills to a different area.** Highlight your ability to search for new scientific information, your communication and other skills, and show how you would be able to adapt these skills to other areas, as the job, course or research requires.

- **Ability to present a set of facts or data to your team.** Tell the interviewer what programs you used, such as graphs generated by Minitab or Excel in your project to display the findings of your study.

After the interview

If you are offered the job or a place on the course, you should accept the offer immediately, unless you have other offers or interviews pending which you see as having a higher priority. Employers will generally allow some time before you accept the offer, but will expect a reply within a reasonable timeframe and may withdraw the offer if you are very slow to reply. Where you have another interview coming up shortly or already have another offer that you are considering, let the prospective employer know. If employers are really keen on you, they may well offer better terms of appointment. Where you have applied for a place on a course, you will find that admissions tutors are likely to be more flexible, unless it is a very popular course. Where you have applied for a research degree, it is sensible to accept the offer: places for research degree are far fewer than those for taught courses.

If you have not heard about the outcome within a couple of days, ring up to enquire. Sometimes a job will have been offered to another candidate who then turns it down. You may be second in line, so do not despair just yet. However, if you do not get the job or a place on the course, ask for feedback as to why you were not chosen. It is best to get information from the chair of the appointing committee or the admissions tutor rather than the secretary. Learn from the feedback and take it into account for your next application and interview. Even a failed interview can be a useful learning process.

Personal development planning

You should have heard of personal development planning (PDP) during your studies and you may have been collecting information to create a portfolio to demonstrate your personal skills acquisition and development. This is currently a requirement of UK degree programmes (as a result of the Dearing Report in 1997). The portfolio you have been compiling should show the skills you have developed to date, including an example of when, how or where you were involved in the relevant activity, such as a first-year undergraduate practical to develop numeracy skills. You may have taken the view that numeracy was a weak area for you and taken steps to complete a short 'catch up' course on mathematics. Later in your degree you will be developing 'graduate skills', including:

- how to retrieve and analyse critically information;
- how to present information;
- how to use information technology.

Skills like these are important to an employer or to an admissions tutor or potential research supervisor because they show you to be the sort of person who can continue to learn, develop and adapt in a changing and demanding job environment.

Some terms

Employers and careers offices may use some terms that you are not familiar with. Table 8.1 gives explanations for some of these terms. It is by no means exhaustive: human resource departments seem to be able to come up with new terms regularly. It is a bit like technical terms in a rapidly developing subject.

Table 8.1 Some terms and their explanation

Term	Explanation
Accessibility	For applications this is used to refer to help in providing documents in alternative formats, such as large print or Braille
Accredited course	A course that is recognized by a professional body and gives formal professional status to the holder
AGCAS	The Association of Graduate Careers Advisory Services. It is the professional body for careers advisors, and publishes helpful material that you may find in your Careers Service centre

Table 8.1 (*Continued*)

Term	Explanation
Border Agency	The body responsible for controlling migration in the UK and enforcing immigration regulations. It looks after any applications from nationals of countries outside the European Economic Area (EEA) and Switzerland for permission to stay, work or study in the UK
Careers fair	An organized event for employers to market themselves by means of exhibition stands that give information about jobs on offer. The fairs can be general or subject specific. They take place in a range of towns and cities, normally where there is a large student population
Data Protection Act	Covers issues of confidentiality particularly of personal data, and the circumstances under which data may be made available to others. An employer or institution offering a course should be registered as a data controller with the Office of the Information Commissioner, and should comply with the regulations under the Act. It will have a statement of the data that will be collected relating to individuals, and the bodies to whom specific data will be disclosed
Equality and diversity	Covers the policy that employers and educational establishments will use when considering applications. All applications have to be considered on the basis of ability, and there should be no discrimination on the grounds of age, gender, disability and so on
Human resources	Human resources departments used to be known as personnel departments or appointment sections. They deal with a range of personnel issues from recruitment to staff support. They will also normally include a section that handles discrimination in its manifest forms
Job specification/person specification	Form part of the overall job description. The job specification lists the detailed requirements, and main duties and responsibilities of the job. The person specification lists the characteristics that are sought for the appointee, such as skills, knowledge and experience
Placement	A short period in employment or voluntary work away from your course, but as part of your degree requirements. It normally forms part of a course which includes 'Applied' in its title. Professional courses normally include some kind of placement
Rehabilitation of Offenders/Criminal Records Bureau (CRB) checks	You are obliged to disclose relevant information about criminal convictions under the Rehabilitation of Offenders Act of 1974. The employer may be obliged to carry out CRB checks. Some jobs may be exempted from the Act and require a CRB check before the individual is appointed. The CRB does have a working code of practice that is available online at: http://www.crb.gov.uk
Shortlist	A list of candidates called for final interview

Further reading

Washington, T. (2004) *Interview Power: Selling Yourself Face to Face.* Mount Vernon Press.

The Government's Career Advice Service gives useful advice on preparing a CV. This will be found at: http://careersadvice.direct.gov.uk/helpwithyourcareer/writecv/?CMP=KAC-jankw08, together with the associated site of CV Builder at: https://www.cvbuilder-advice-resources.co.uk/careersadvice/index.php

Index

Locators in **bold** refer to major content
Locators in *italic* refer to figures/tables
Locators for main headings which also have subheadings refer to general aspects of the topic

---

Index by Judith Reading